Fish Genetics and Breeding

The Authors

T L S Samuel Moses is currently working as Assistant Professor in Department of Aquaculture, Tamil Nadu Dr. M.G.R. Fisheries College and Research Institute, Tamil Nadu Dr. J. Jayalalithaa Fisheries University, Ponneri. He is having experimental knowledge in the field of fish genetics and breeding for 7 years. He is currently handling courses at UG & PG level in the specialization of fish genetics and breeding. He has published 7 research papers, 4 teaching/training manuals and organized 5 training programmes. He has completed 1 University research project and also involved in 3 externally funded research projects.

Dr S Felix, Ph.D., Vice Chancellor, Tamil Nadu Dr. J. Jayalalithaa Fisheries University, Nagapattinam is having experience of more than 35 years in the area of fisheries and aquaculture in teaching, research, extension and administration. He has organized more than 20 Nos. of seminar/conference/symposium at National and International levels. Currently he is also the President (2018-19) of World Aquaculture Society, Asian Pacific Chapter and Chairman, ICAR's BSMA Committee (2018-19). He has more than 60 research papers published in National and International journals. He has authored more than 10 books and 25 manuals. He is specialized in the area of advanced systems in aquaculture and aquariculture such as raceway, biofloc technology, RAS, etc,. He has operated around 20 externally funded research projects.

Fish Genetics and Breeding

by

T L S Samuel Moses

S Felix

Tamil Nadu Dr. M.G.R. Fisheries College & Research Institute
Tamil Nadu Dr. J. Jayalalithaa Fisheries University
Ponneri, Tiruvallur District, Tamil Nadu

DAYA PUBLISHING HOUSE®
A Division of
ASTRAL INTERNATIONAL PVT. LTD.
New Delhi – 110 002

ISBN: 9789388173223 (Int. Edition)

Published by	:	**Daya Publishing House®** *A Division of* **Astral International Pvt. Ltd.** – ISO 9001:2015 Certified Company – 4736/23, Ansari Road, Darya Ganj New Delhi-110 002 Ph. 011-43549197, 23278134 E-mail: info@astralint.com Website: www.astralint.com
Laser Typesetting	:	**Classic Computer Services,** Delhi - 110 035
Printed at	:	**Neelam Graphics, Delhi - 110007**

डॉ. जे. के. जेना
उप महानिदेशक (मत्स्य विज्ञान)

Dr. J. K. Jena
Deputy Director General (Fisheries Science)

भारतीय कृषि अनुसंधान परिषद
कृषि अनुसंधान भवन-II, पूसा, नई दिल्ली 110 012
INDIAN COUNCIL OF AGRICULTURAL RESEARCH
KRISHI ANUSANDHAN BHAVAN-II, PUSA, NEW DELHI - 110 012

Ph. : 91-11-25846738 (O), Fax : 91-11-25841955
E-mail: ddgfs.icar@gov.in

Foreword

Quality seed remains the most important critical input for any farming operation, so also in aquaculture. Aquaculture, which has witnessed significant transformation over the years and is looking for greater diversification, today is constrained with non-availability of quality seed of desired species. Further, the genetic quality deterioration in several cultured fish species has been a concern for all associated with the fisheries sector. Generally, genetics research is undertaken to develop high-quality strains in the species of commercial importance, in order to improve the production performance and market value of the produce. The basic principle involved in breeding is to select the superior strains and establishing the superior characteristics of the parents in the offspring. Both short-term and long-term genetics approaches are used in the improvement programmes, which are generally taken up based on the species-specific requirements, resource availability and the timeline for obtaining the desired advantage. With captive breeding of aquatic species becoming easier & more farmers'-friendly and increasing number of species domesticated, the risk of mixing of these stocks with their wild relatives in the natural environment has increased substantially, thereby requiring appropriate strategies for effective management of the natural resources.

The book has adequately covered various fundamental and applied aspects of genetics, including monohybrid & dihybrid cross, heterosis in fish breeding, sex determination, inbreeding, selection and heritability and gene and genotype frequency, protocol for induction of triploidy and gynogenesis in fishes, cryopreservation of fish sperm and several other relevant subjects.

I appreciate the effort of the authors Drs. T.L.S. Samuel Moses and S. Felix in bringing out this useful publication.

(J.K. Jena)

Message

Aquaculture is the fastest growing food producing sector in the world. Seed is one of the basic input for the fish production. Even though the breeding and hatchery production of seeds had been standardized for number of candidate species for aquaculture, quality seed is the need of the hour. Knowledge on inbreeding and heterosis will lead to better understanding of involvement of genetic principles in fish production. The problems worked out in various chapters will give the in-depth practical application of genetics principles in sustainable increase of fish production.

S. Felix
Vice Chancellor
Tamil Nadu Dr. J. Jayalalithaa Fisheries University

Preface

This practical manual on "Fish Genetics and Breeding" was prepared as per the ICAR syllabus for UG students and will cover basics to modern concepts.

Various problems on concepts based on mendalian genetics such as Monohybrid cross, incompletely dominant genes, Lethal genes, Dihybrid cross are given with solution and answers.

Detailed protocol with explanation is also given for induction of Triploidy in fishes, Induction of Gynogenesis in fishes, Preparation of metaphase chromosomes from fish tissues, chromosome preparation from embryos, Cryopreservation of fish sperm and checking the fish sperm motility

Problems based on other important concepts in fish breeding such as Heterosis in fish breeding, sex determination, inbreeding and effective breeding number, selection and heritability and gene and genotype frequency are also given along with suitable solution and answers.

We hope this manual will be much of use to the students and fisheries professionals.

Authors

Contents

Chapter 1

Problems on Monohybrid Cross

Introduction

Monohybrid Cross

When a cross is made between two parents differing in a single character, the cross is known as monohybrid cross.

Back Cross

When the offspring (F_1 individuals) are crossed with one of the two parents then such a cross is called as **Back cross**.

When the offspring (F_1 individuals) is back crossed with its recessive parent, it is called **Test cross.** Test cross establishes the homozygosity/Heterozygosity of the offspring. If the offsprings are heterozygous a ratio of 1:1 (dominant: recessive) will be expressed during the test cross. If the offsprings are homozygous only the dominant trait will be expressed.

Dominant and Recessive

In a heterozygote only one allele of the pair is able to express itself, while the other is not able to do so. The former is known as **dominant allele** and the latter as **recessive allele**.

The dominant allele expresses itself in the homozygous and heterozygous condition.

A phenotypic trait or character which appears only in the homozygous individual is called a recessive trait and the pair of alleles which specifies a recessive phenotypic trait is called a recessive allele.

While a phenotypic trait or character which appears both in the homozygous as well as in heterozygous individual is called a dominant trait and the pair of alleles which specifies a dominant phenotypic trait is called a dominant allele.

Genotype

The genetic make-up of each fish is called its "genotype". Because chromosomes occur as pairs, each gene occurs as a pair; this means the genotype is a paired entity.

Phenotype

Any measurable characteristic or distinctive trait possessed by an organism is called **phenotype** of that organism. Each phenotypic trait of an organism is caused and determined by genotype and environmental condition.

Problem No. 1

A dominant allele **G** produces grey guppies. Its recessive allele **g** produces gold guppies. A group of homozygous grey guppies and gold guppies were crossed and their F_1 progeny were then test crossed. Determine the expected genotypic and phenotypic ratios among F^1 and F^2 progeny.

Solution

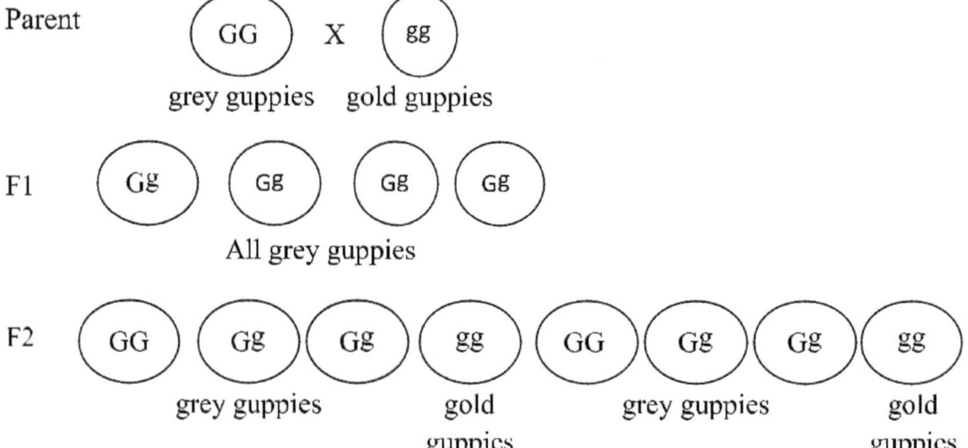

Answer

1. Genotypic and phenotypic ratio among F^1 and F^2 progeny are as follows:
2. F^1 All Gg and all grey guppies

   F^2 GG:Gg:gg (1:2:1) and Grey guppies: Gold guppies (3:1)

Problem No. 2

In the case of goldfish, the cross of normal eye X telescopic eye yields normal and telescopic eyed progeny in equal proportions, but telescopic eyed X telescopic eyed always gives rise to only the telescopic eyed progeny. What does this tell about the genotype of normal and telescopic eyes in goldfish? Which phenotype is dominant?

Solution

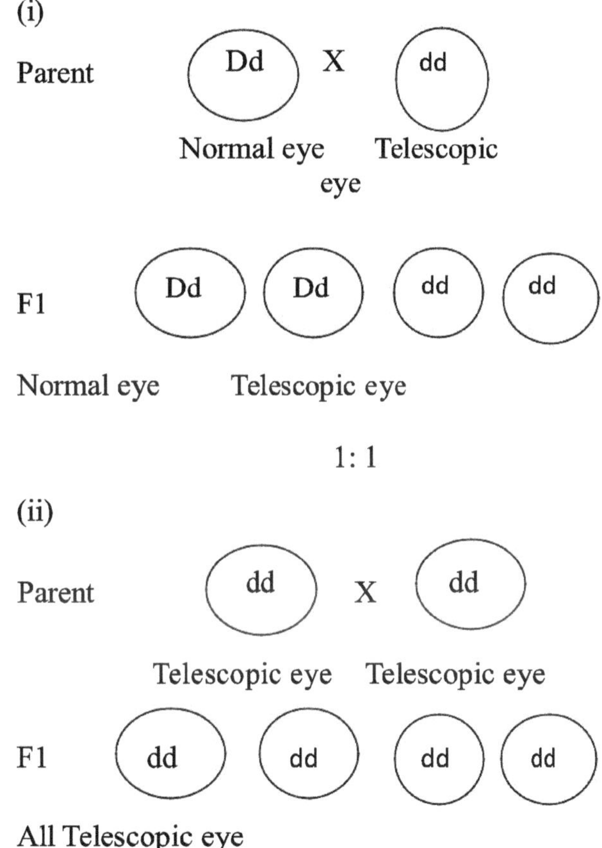

Answers

From the above it is understood that,

- ☆ 'D'- Normal eyes – dominant
- ☆ 'd'- Telescopic – recessive
- ☆ Genotype of normal eyes – DD/Dd
- ☆ Genotype of telescopic eyes – dd

Therefore, normal eyes are dominant.

Problem No. 3

The colour of the parent and offsprings of grass carp is given below. Explain the results for the crosses and mention the genotypes of all the individuals.

Sl.No.	Parents	Progeny Colouration
1.	Wild x Wild	Wild
2.	Wild x Albino	Wild
3.	Albino x Wild	Wild
4.	Albino x Albino	Albino

Solution

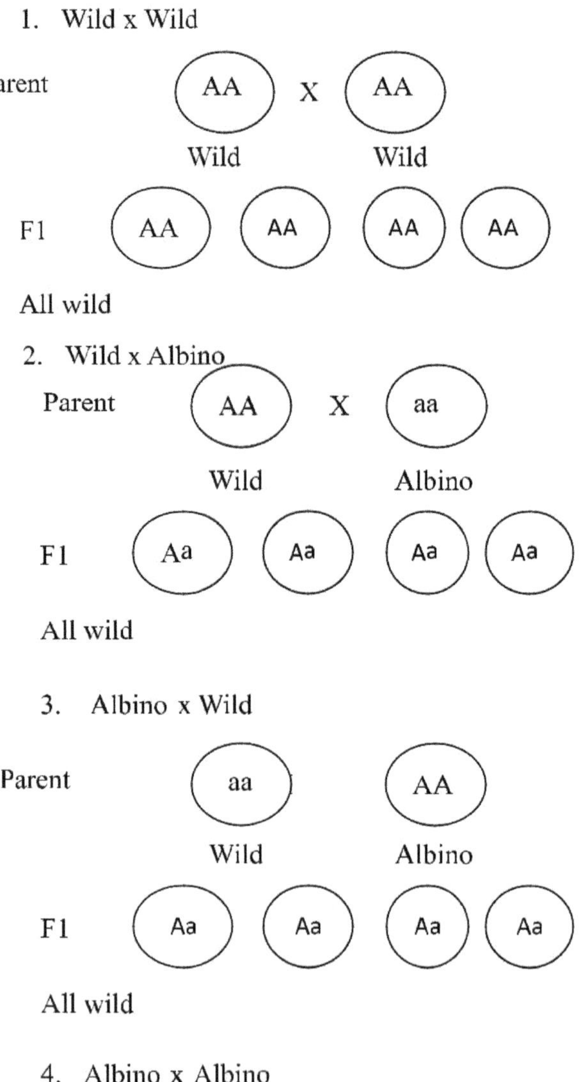

4. Albino x Albino

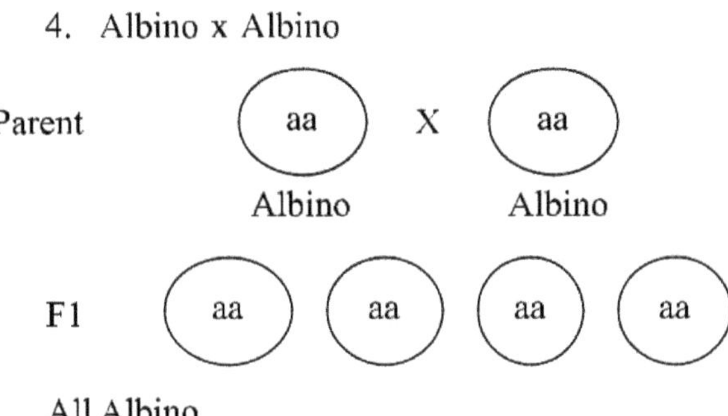

Answers

i) Explanation

In the above problem the wild type is the dominant gene, hence it can express itself both in homozygous and heterozygous state. So, the cross of Wild x Wild, Wild x Albino and Albino x Wild always yields only wild. As the albino type is a recessive gene the cross of Albino x Albino gives rise to all albino progeny.

ii) Genotype

Sl.No.	*Parents*		*Progeny Coloration*
1.	Wild x Wild		Wild
	AA	AA	AA
2.	Wild x Albino		Wild
	AA	aa	AA
3.	Albino x Wild		Wild
	aa	AA	AA
4.	Albino x Albino		Albino
	aaaa		aa

Problem No. 4

Normal eyes of gold fish is a dominant trait. Telescopic eyes is the alternative recessive trait. When a pure normal eye gold fish is crossed to a telescopic one,what fraction of F_2 is expected to be heterozygous?

Solution

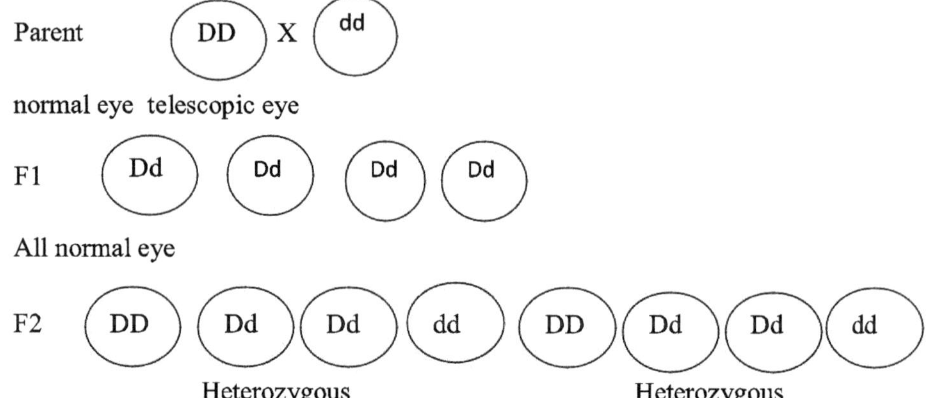

Answer

50 per cent of the F2 is Heterozygous

Problem No. 5

In carp, blue colour is recessive to non blue colour. What coloured carp will you choose to breed a given non blue colour fish in order to find its genotype? Write the type of cross?

Solution

- ☆ Non blue colour – dominant
- ☆ Blue colour – recessive

i) Heterozygous non-blue parent

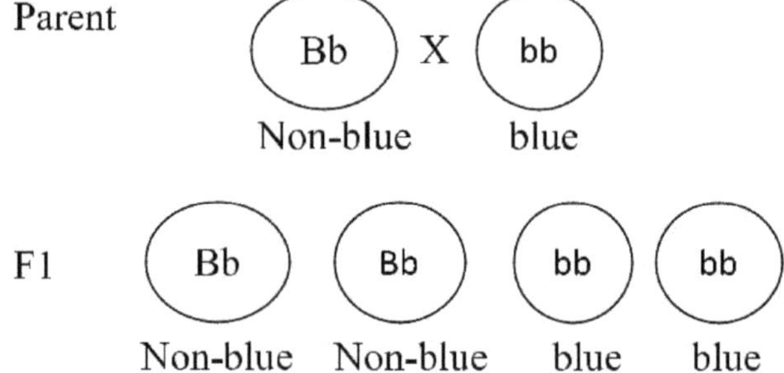

ii) Homozygous non-blue parent

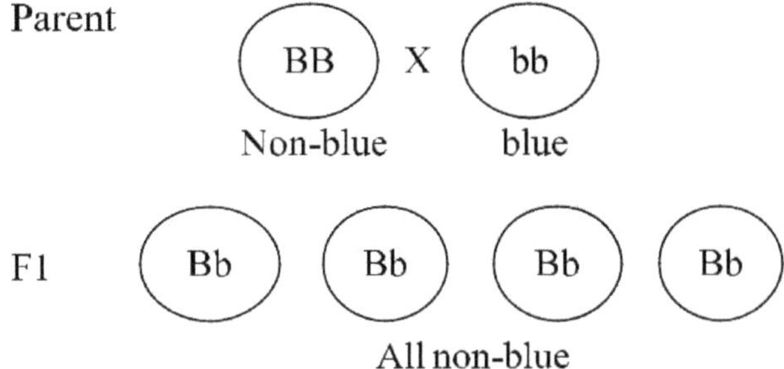

Answer

Blue colour carp will be chosen and type of cross is test cross. As test cross gives 1:1 ratio when heterozygous non-blue carp is used as parent and gives all non-blue carp when homozygous non-blue carp is used as parent.

Problem No.6

Koi carps with known phenotype but with unknown genotype produced the listed progeny.

Sl.No.	*Parents*	*Offsprings*	
		Pattern	
1.	Pattern x No pattern	820	780
2.	Pattern x Pattern	1180	390
3.	No pattern x No pattern	0	1500
4.	Pattern x No pattern	3500	0
5.	Pattern x Pattern	2800	0

Using **D** for the light yellow pattern gene and **d** for no pattern give the most probable genotype of each parent.

Solution

The most probable ratio of the above parent are calculated as follows:

Sl.No.	*Parents*	*Offsprings*		*Ratio*
		Pattern	*No Pattern*	
1.	Pattern x No pattern	820	780	1:1
2.	Pattern x Pattern	1180	390	3:1
3.	No pattern x No pattern	0	1500	All no pattern
4.	Pattern x No pattern	3500	0	All pattern
5.	Pattern x Pattern	2800	0	All pattern

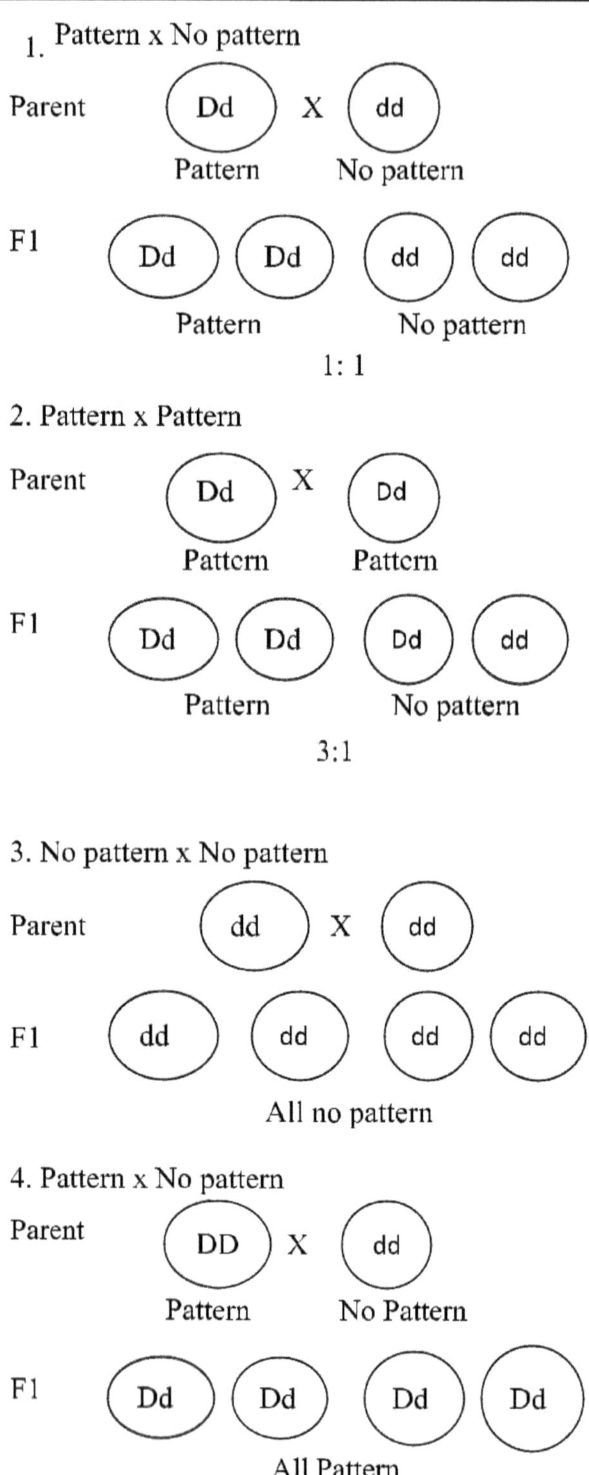

5. Pattern x Pattern

5. Pattern x Pattern

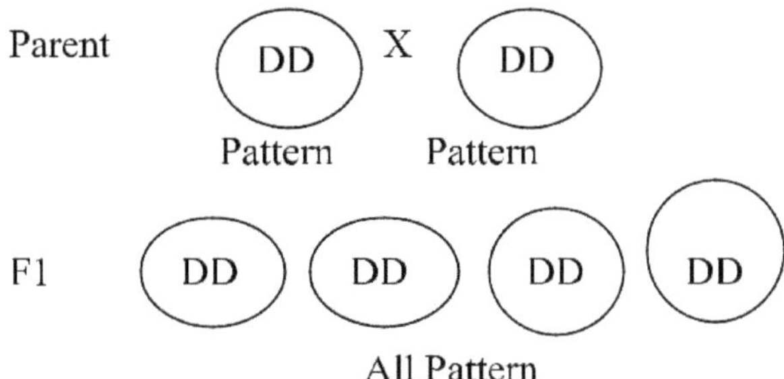

Answers

The most probable genotype of the parents are as follows:

Sl.No.	Parents	Most Probable Genotype
1.	Pattern x No pattern	Dd x dd
2.	Pattern x Pattern	Dd x Dd
3.	No pattern x No pattern	dd x dd
4.	Pattern x No pattern	DD x dd
5.	Pattern x Pattern	DD x DD

Problem No.7

In carps blue colour is recessive to non-blue colour. A non blue colour carp male has blue colour female parent. What is the genotype of the fish and female parent? What are the possible genotypes of male parent? If the fish is crossed with a blue coloured female, what are the possible genotypes of their offsprings?

Solution

Two type of genotype of non-blue colour male carp is possible as below.

(i) Homozygous non-blue male carp

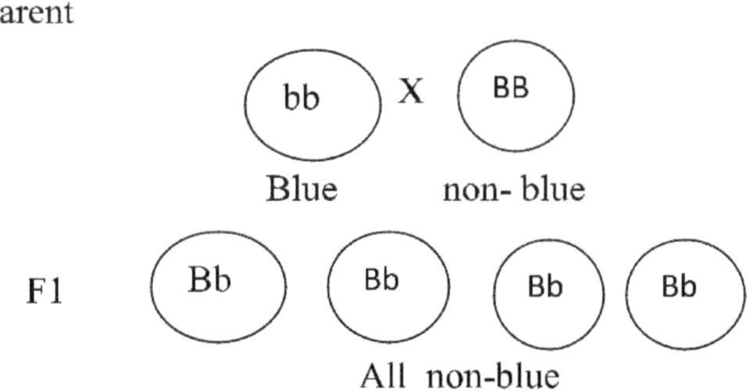

ii) Heterozygous non-blue male carp

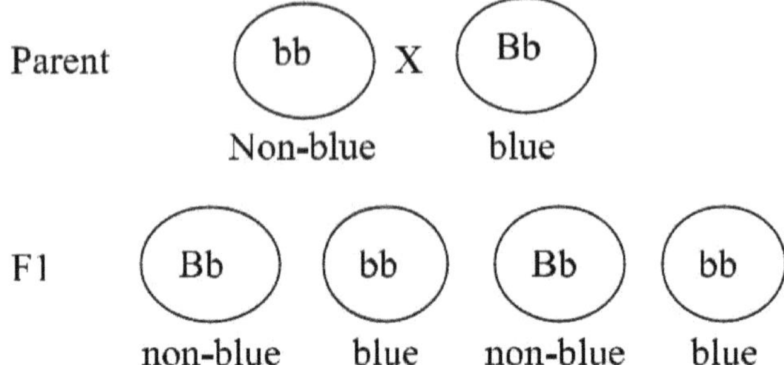

Answers

- ☆ The genotype of blue female carp parent is bb
- ☆ The genotype of non-blue male parent is BB/Bb
- ☆ The genotype of offspring in case of homozygous non-blue male carp is all Bb (Non- blue)
- ☆ The genotype of offspring in case of heterozygous non-blue male carp is Bb:bb (Non-blue: blue) 1:1

Problem No. 8

Two light yellow pattern Japanese carps on breeding produce 1678 light yellow pattern and 554 no pattern offsprings. Give the genotypes of the parents and justify your answers with reasons. (Given: Light yellow pattern is dominant to no pattern). If a pure light yellow pattern carp is crossed with a hybrid light yellow pattern, what will be the genotypic ratio of the offsprings?

Solution

- ☆ D-Light yellow pattern is dominant
- ☆ d - no pattern is recessive

Two light yellow pattern Japanese carps on breeding produce 1678 light yellow pattern and 554 no pattern offsprings. which gives 3:1 ratio and hence the parent are heterozygous light yellow pattern Japanese carps.

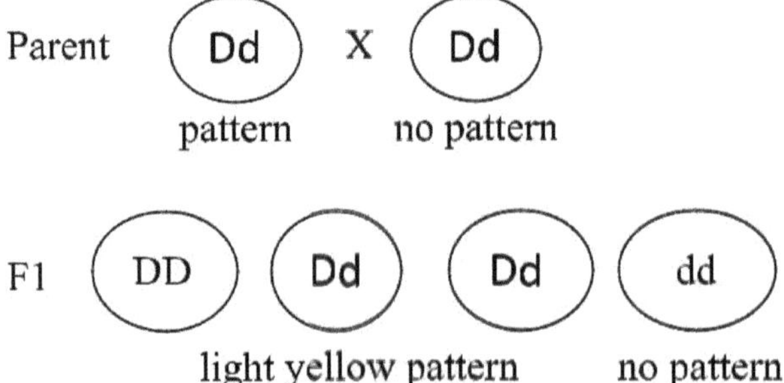

☆ DD - pure light yellow pattern carp

☆ Dd - hybrid light yellow pattern carp

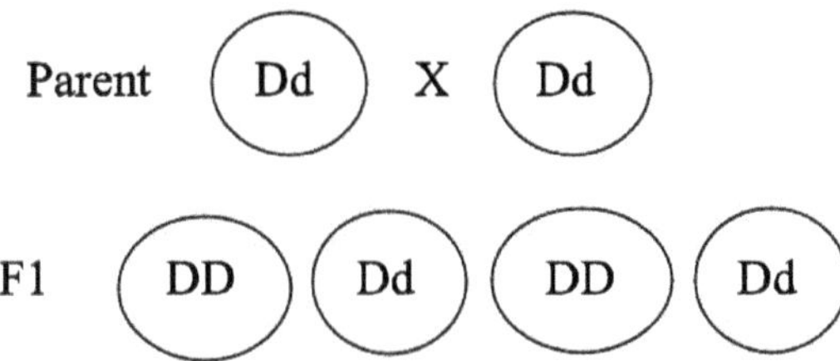

1 : 1 (light yellow pattern carp: hybrid light yellow pattern)

Answers

1. Genotypes of the parents – Dd x Dd
2. When a pure light yellow pattern carp is crossed with a hybrid light yellow pattern the genotypic ratio of the offsprings will be of 1: 1 (homozygous light yellow pattern: hybrid yellow pattern)

Problem No. 9

When a red Medaka was crossed with wild type, the offspring were all wild type (green), but when these offspring were crossed among themselves the result was approximately three greenish offspring to one red. Explain dominant and recessive colour in Medaka. What is the type of inheritance in Medaka?

Solution

'G' wild type Medaka is dominant gene then the recessive gene is 'g' for red Medaka, as the cross between the F^1 offsprings gives 3:1 ratio and all the F^1 offspring are wild type, the parent cross should be between pure wild type and red Medaka as given below:

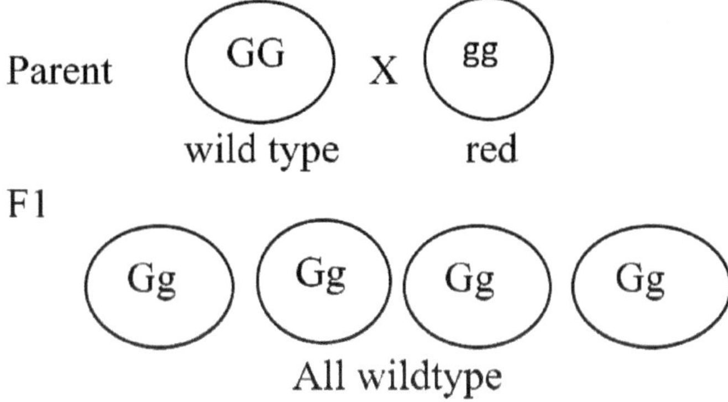

When the offspring F1 were crossed among themselves, it results in 3: 1 (wild type : red) ratio as given

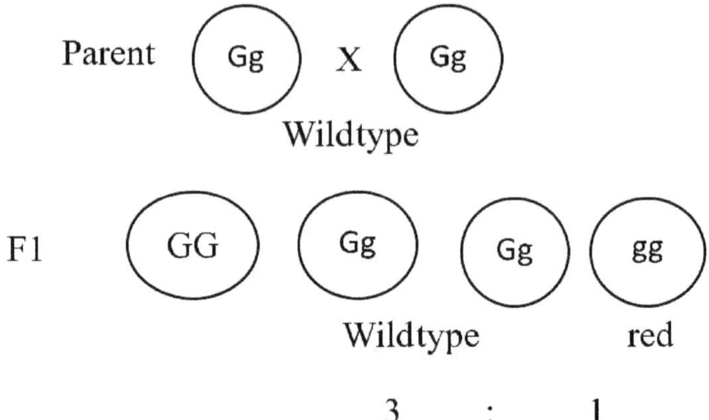

Answers

- ☆ Green (wild) is dominant and red is recessive.
- ☆ Mendelian pattern of inheritance is complete dominance.

Problem No. 10

A dominant gene '**+**' is responsible for the normal pigmentation of body colour in catfish; its recessive allele '**a**'produces albino. A test cross of a normal type female gave 5200 normal and 5500 albino in the F_1. If the normal F_1 females are crossed to their albino F_1 brothers, What genotypic and phenotypic ratios would be expected in the F_2? Diagram the results using the appropriate genetic symbols.

Solution

A test cross of a normal type female gave (5200 normal and 5500 albino) 1:1 ratio in the F_1 Therefore parent are + a x aa the normal F_1 females are crossed to their albino F_1 brothers as given below

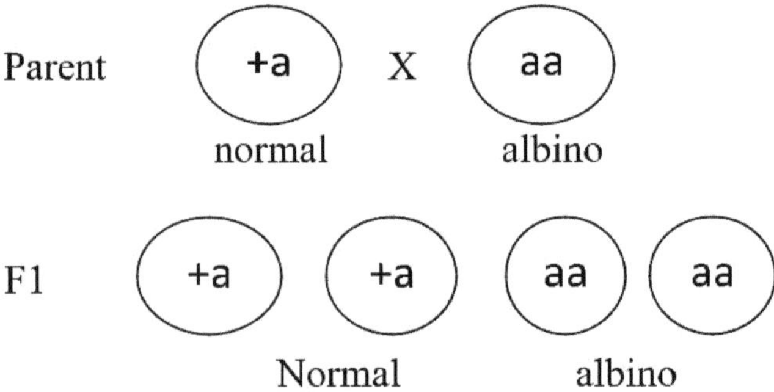

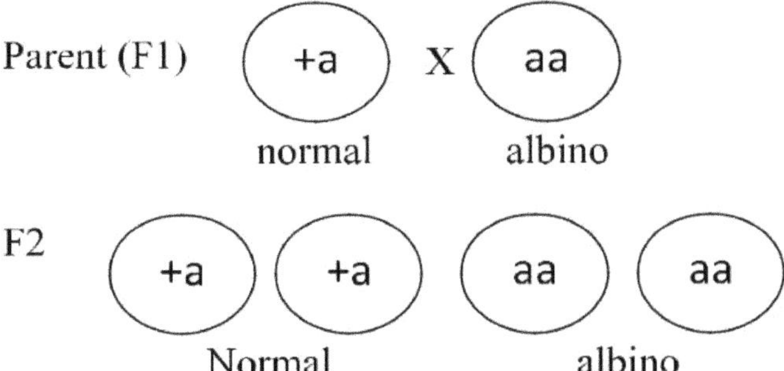

Answers

- ☆ Since the recessive albino phenotype appears in the F_1 in approximately a 1:1 ratio, we know that the female parent must be heterozygous +a. Furthermore, we know that the normal type F_1 progeny must also be heterozygous +a.
- ☆ The normal type F_1 females are then crossed with their albino brothers which gives the following ratio:

  F_1cross: +a ♀ x aa ♂ (normal x albino)

  F_2 +a:aa = 1:1 (normal x albino)

Chapter 2

Problems on Incompletely Dominant Genes

Introduction

This type of dominance occurs when the dominant allele is **incompletely dominant**, *i.e.*, the dominant allele always produces its phenotype, but it is unable to complete suppress of the recessive allele in the heterozygous state. When this happens, a gene with incomplete dominant gene action (it has an incompletely dominant allele and a recessive allele) has three genotypes and each genotype produces a unique phenotype.

The heterozygous genotype produces a phenotype that resembles but is slightly different from the dominant phenotype. Because of this, the dominant phenotype can be produced only when a fish has two copies of the dominant allele (homozygous dominant). Since the recessive allele is not completely suppressed by the dominant allele, the heterozygous genotype produces a phenotype that resembles, but is not identical to the dominant phenotype.

Problem No. 1

In red tilapia, the allele for red body colour has an effect that is incompletely dominant over grey or black colour allele. If a cross between two tilapias produced 32 red, 68 pink and 28 black tilapias, what are the phenotypes of the parents?

Solution

As the cross between two tilapias produced 32 red, 68 pink and 28 black tilapias which roughly works out to 1:2:1 (red: pink: black tilapias), this ratio results only when two heterozygous genotypes are cross as follows:

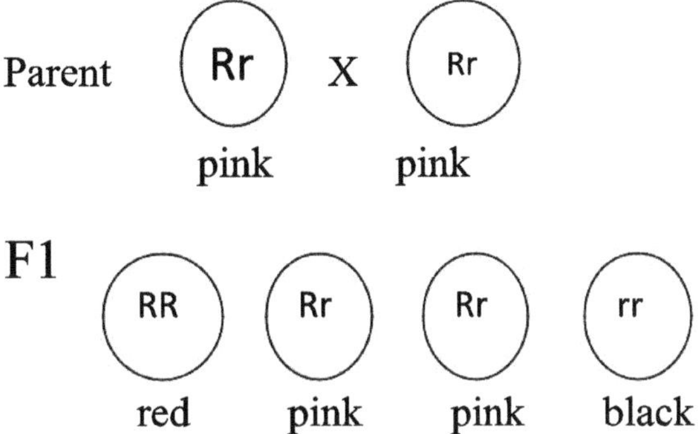

1:2:1 (red: pink: black tilapias)

Answer

Phenotypes of the parents: Pink x Pink

Problem No. 2

Calculate the ratios of body colour in red tilapia among the offsprings produced for the following crosses?

- ☆ red x red
- ☆ red x pink
- ☆ pink x black
- ☆ pink x pink

Solution

a) red x red

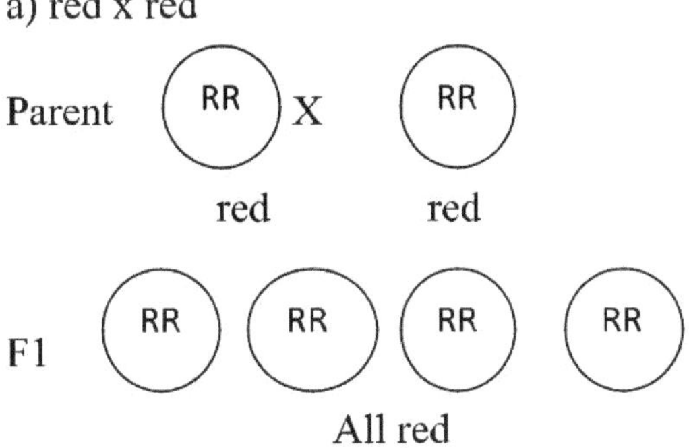

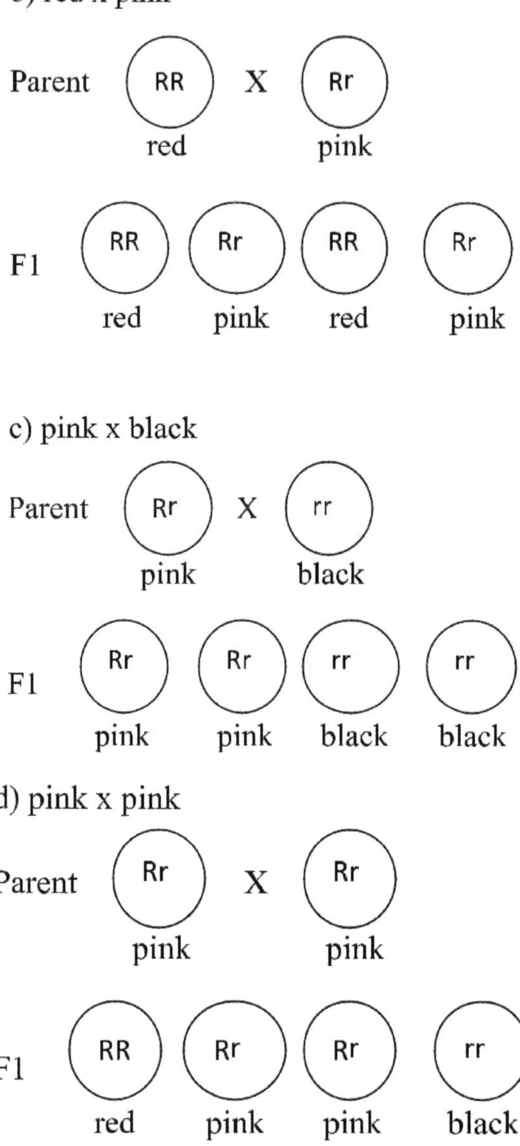

Answers

The ratio of body colour in red tilapia among the offsprings produced for the above crosses is as follows:

- All red
- Red: Pink (1:1)
- Pink: Black (1:1)
- Red:Pink: Black (1:2:1)

Problem No. 3

When a transparent scaled (T') goldfish is crossed to a normal scaled goldfish, all the progeny are calico type (T T'). From a cross of two such F_1s repeatedly made, the F_2 showed 5000 transparent scaled, 10000 calico and 5000 normal scaled goldfishes. How this trait is inherited? Give the genotypes of the parents, F_1 and F_2 progeny.

Solution

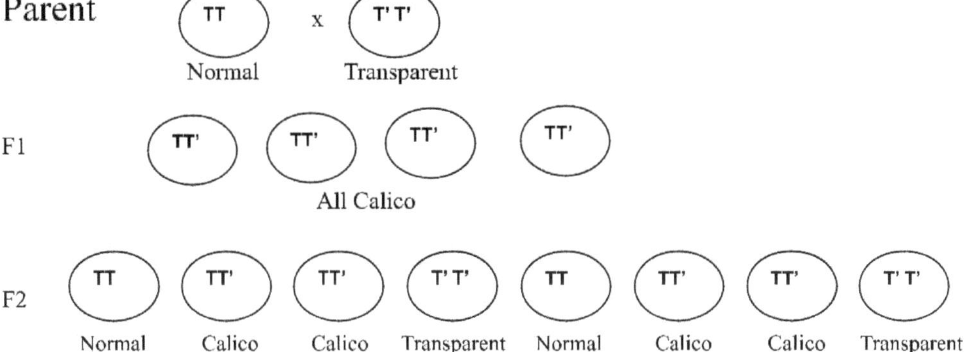

Answers

i) The inheritance of scale pattern in goldfish is due to incompletely dominant gene action.
   - ☆ Genotypes of the parents: T1T1 x TT
   - ☆ Genotypes of the F_1: All T1T
   - ☆ Genotypes of the F_2: T1T1:T1T: TT (1:2:1)

Problem No. 4

Give the genotypic and phenotypic ratio for the offsprings produced for the following crosses in the Siamese fighting fish.

- ☆ steel blue x steel blue
- ☆ steel blue x blue
- ☆ steel blue x green
- ☆ blue x blue
- ☆ blue x green
- ☆ green x green

Solution

a) steel blue x steel blue
Parent
VV
X
VV
steel blue
steel blue
F1
VV
VV
VV
VV
All steel blue
b) steel blue x blue
Parent
VV
X
Vv
steel blue
blue
F1
VV
Vv
VV
Vv
steel blue
blue
steel blue
blue
c) steel blue x green
Parent
VV
X
vv
steel blue
green
F1
Vv
Vv
Vv
Vv
All blue
d) blue x blue
Parent
Vv
X
Vv
blue
blue
F1
VV
Vv
Vv
vv
steel blue
blue
blue
green

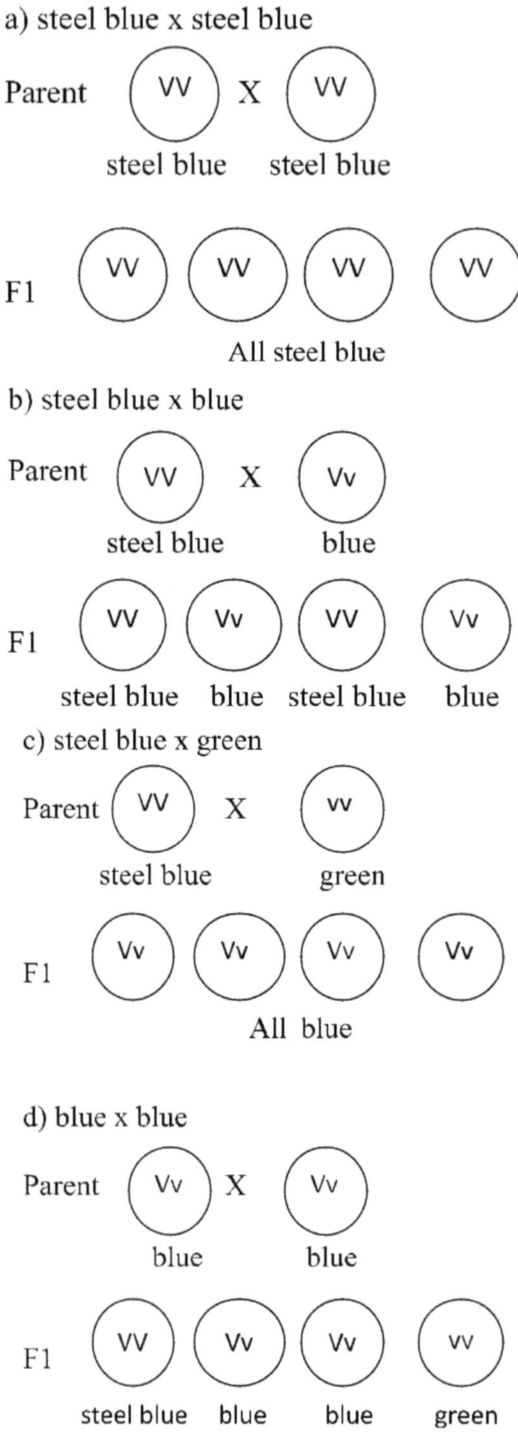

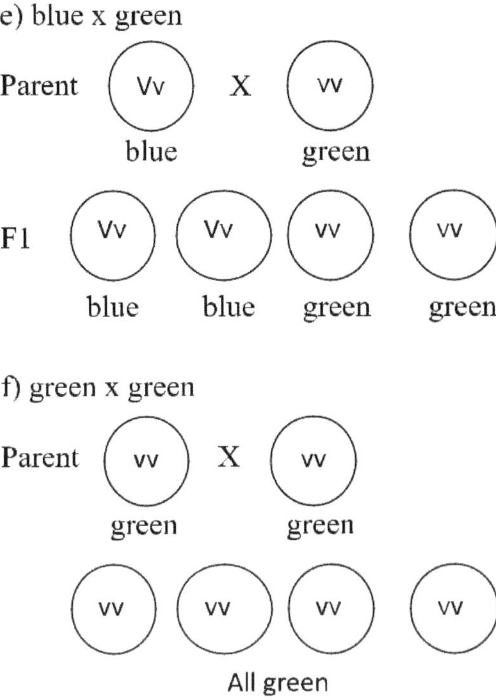

Answers

- ☆ All steel blue (VV)
- ☆ Steel Blue: Blue (VV: Vv, 1:1)
- ☆ All blue (Vv)
- ☆ Steel blue:blue: green (VV:Vv: vv, 1:2:1)
- ☆ Blue: Green (Vv: vv, 1:1)
- ☆ All green (vv)

Problem No. 5

Three alleles determine the ABO blood type in brown trout, *Salmo trutta fario*. Alleles I**a** I**a** produces the A phenotype, I**b** I**b** produces the B phenotype, and I**o** I**o** produces O phenotype. The effects of I**a** and I**b** are dominant to O but are co dominant with each other. What are the genotypes of the following parents?

Sl.No.	*Phenotypes of Parents*	*Phenotypes of Offspring Proportions*			
		A	*B*	*AB*	*O*
1.	B x B	-	¾	-	¼
2.	B x AB	-	½	½	-
3.	B x A	-	½	½	-

Sl.No.	Phenotypes of Parents	Phenotypes of Offspring Proportions			
		A	B	AB	O
4.	B x A	¼	¼	¼	¼
5.	B x O	-	1	-	-
6.	B x AB	¼	½	¼	-

Solution

1. B x B

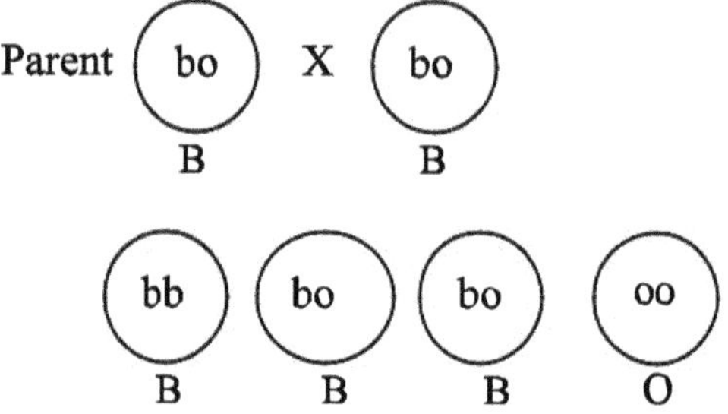

2. B x AB

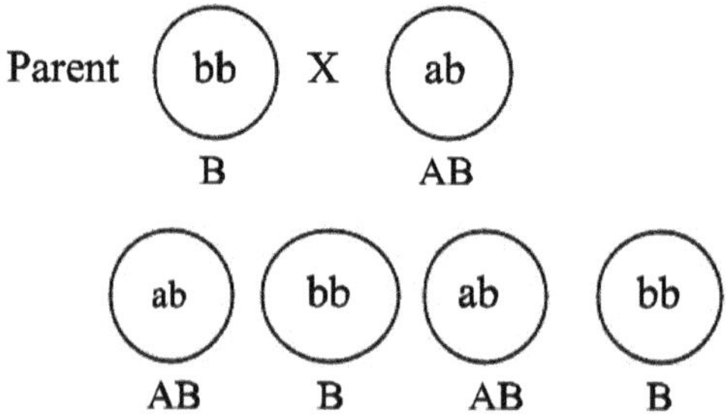

3. B x A

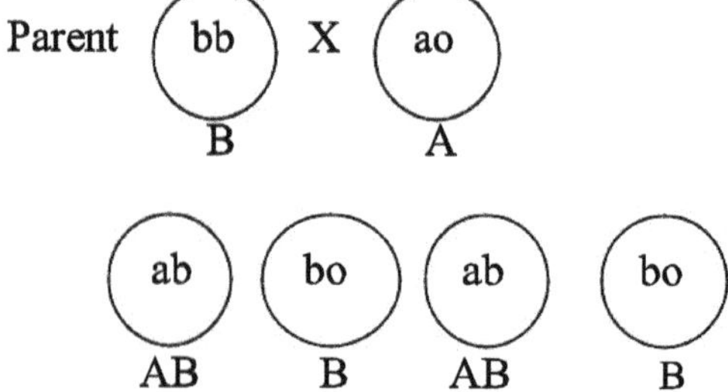

- B x A

Parent bo X ao

B A

ab bo ao oo

AB B A O

5. B x O

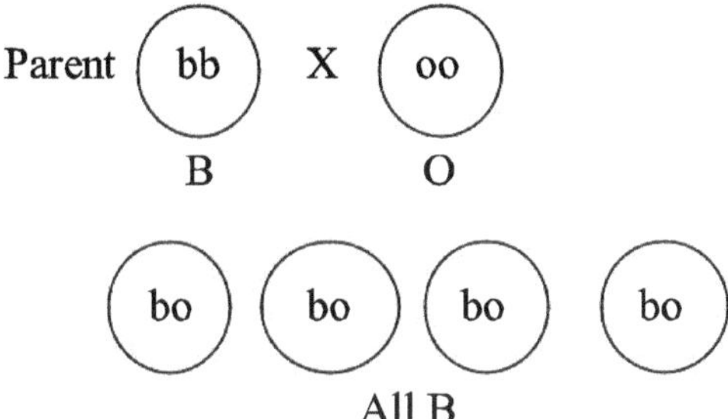

6. B x AB

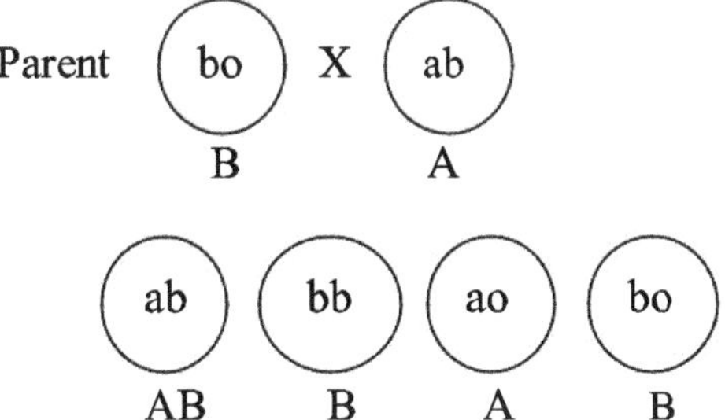

Answers

The genotypes of the parents are as follows:

- ☆ bo x bo
- ☆ bb x ab
- ☆ bb x ao
- ☆ bo x ao
- ☆ bb x oo

Problem No. 6

In Tilapia, gold individuals were shown to be homozygous for the recessive allele (GG), while normal (wild) coloured individuals were homozygous for the normal (WW) allele (black colour) expression. Hetrozygotes (WG) displayed a "bronze" skin colour. Give the genotypic and phenotypic ratios for the offsprings produced among the following cross.

- ☆ GG x WW
- ☆ WG x WG
- ☆ WG x WW
- ☆ WG x GG

Solution

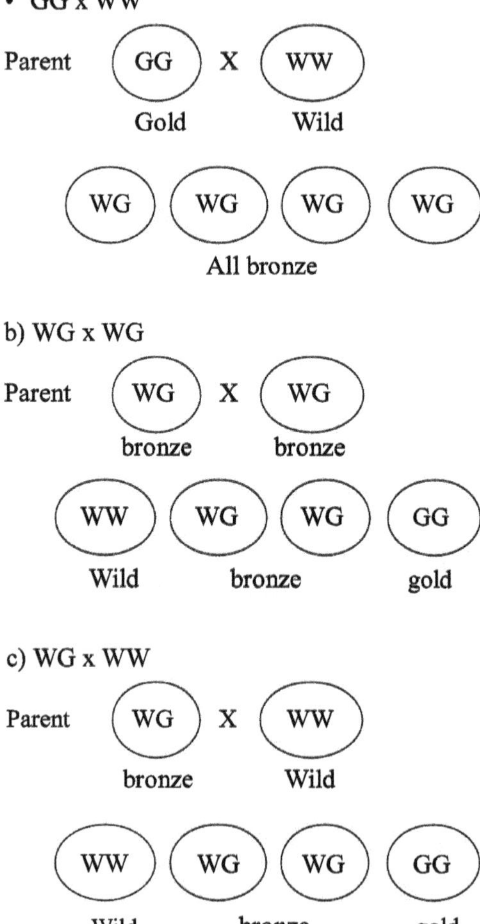

d) WG x GG

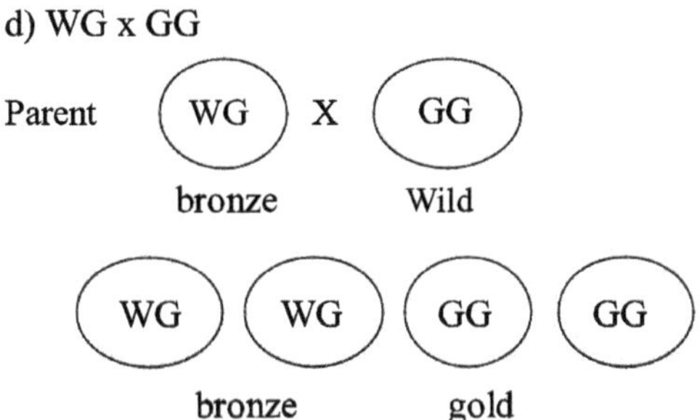

Answers

1. All Bronze
2. 1:2:1
3. Wild: Bronze (1:1)
4. Bronze: Gold (1:1)

Chapter 3

Problems on Lethal Genes

Introduction

A gene whose phenotypic effect is sufficiently drastic to kill the bearer is called **lethal gene**. Death from different lethal genes may occur at any time from fertilization of the egg to advanced age. Lethal genes may be dominant, incompletely dominant or recessive. The fully dominant lethal gene kills the carrier individual both in homozygous and heterozygous conditions.

Lethal Genes

A gene whose phenotypic effect is sufficiently drastic to kill the bearer is called **lethal gene**. Death from different lethal genes may occur at any time from fertilization of the egg to advanced age.

1. Lethal genes may be dominant, incompletely dominant or recessive.
2. The fully dominant lethal gene kills the carrier individual both in homozygous and heterozygous conditions.
3. Completely dominant lethal genes usually cause death of the zygote later in embryonic development or even after birth or hatching. No individuals will attain the age of reproduction.
4. Lethal genes arises occasionally by mutation from normal allele. The individual with a dominant allele die before they can produce the progeny. Therefore the mutant dominant allele is removed from the population in the same generation in which it arose.
5. The recessive lethal allele kills the carrier individual only in homozygous condition.

6. Certain lethal genes produce certain disorder to the individuals but do not destroy the individual completely. They handicap but do not destroy. Such genes are called **semi lethal genes** or **sub lethal genes** or **subvital genes**.

Dominant Lethal Genes

Dominant lethal gene in common carp.

1. Lethal genes **N** (reduction of scales) and **L** (lighter pigmentation) kill the carriers in the homozygous state.
2. The S gene in *Tilapia aurea* in another example of a dominant lethal gene.

Genotype	*Phenotype*
SS	Death
S+	Saddleback (abnormal dorsal fin)
i	Normal

The mating of two saddleback **(S+)** *T. aurea* produces the expected 1:2:1 genotypic and phenotypic ratios, but because all homozygous dominant fish are aborted, the genotypic ratio 2S+:1 ++ and 2 saddleback: 1 normal phenotypic ratios are observed.

Recessive Lethal Genes

1. In guppy, *Poecilia reticulata*, when two Y-chromosomes are combined in a male, homozygotes with respect to genes **ma** (maculatus), **ar** (Armatus) and **pa** (Pauper) turn out to be non-viable.
2. Meanwhile, the males with genotypes **YmaYar**, **YmaYpa** and **YpaYar** are viable and in addition fertile.
3. It appears that a special lethal gene is closely linked with each of the genes located in Y chromosome;
4. The lethal genes associated with different genes are non-allelic.

Semi-Dominant Lethal Gene

1. The locus **N** (Nigra) in platy is a result of dominant mutation and is widely used by aquarium fish culturists. The mutation results in a markedly blackening of the tail part of the body and of the caudal fin. The pigmentation is due to the accumulation of large melanophores.
2. The semi dominant gene **Fu** producing the black pigmentation of the whole body appears to be an allele of the gene **N**.
3. The homozygotes **Fu** are non-viable. In the presence of recessive allele **n** the macromelanophores are completely absent.

4. In angel fish, *Pterophyllum scalare,* there are about 25 colour patterns which are the most important ornamental traits in this species. Golden body colour is determined by a recessive autosomal gene designated by a semi-lethal autosomal gene, **BI**.

Sex-Linked Lethal Gene

The X and Y chromosome of medaka, *Oryziaslatipes* contain one "pigment" locus R with three alleles.

- ☆ Normal fishes living in nature contain R gene in the sex chromosomes X and Y.
- ☆ Genetic analysis has shown that YR YR males obtained in crossing are practically non-viable and the lethal ratio 2:1 has been reported.

Problem No. 1

The **S** gene in *Tilapia aurea* is an example of a dominant lethal gene. The gene **S+** produces saddleback and the gene **++** produces normal fin. Find out the genotypic and phenotypic ratios for the following crosses.

- ☆ Saddleback x Saddleback
- ☆ Saddleback x Normal
- ☆ Normal x Normal

Solution

a) Saddleback x Saddleback

Parent **S+** X **S+**

Saddleback Saddleback

F1 **SS** **S+** **S+** **++**

Lethal Saddleback normal

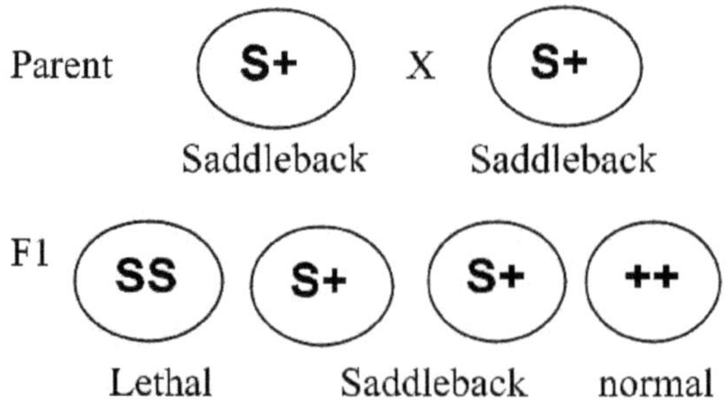

b) Saddleback x Normal

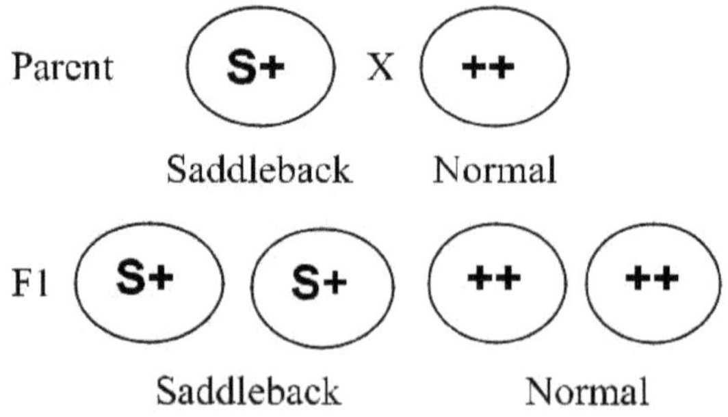

Parent **S+** X **++**

Saddleback Normal

F1 **S+** **S+** **++** **++**

Saddleback Normal

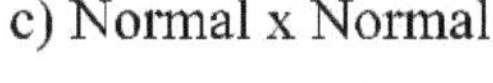
c) Normal x Normal

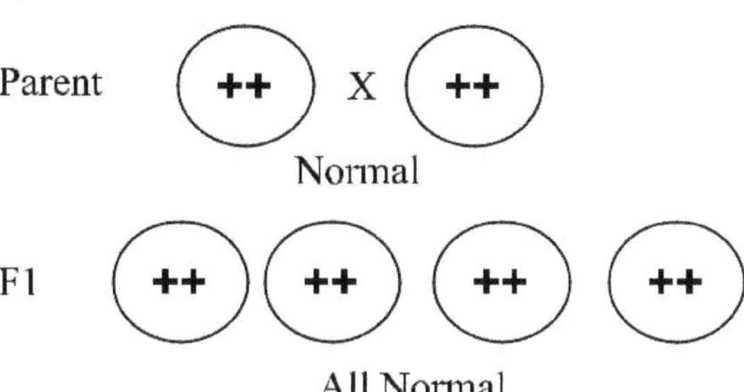

Answers

a) Genotypic ratio

☆ : S+: + + = 1:2:1 Phenotypic ratio

lethal: saddle back: normal= 1:2:1

b) Genotypic ratio

S+:+ + = 1:1

Phenotypic ratio

saddleback: normal =1:1

c) Genotypic ratio

All ++

Phenotypic ratio

All normal

Problem No. 2

In common carp, a dominant gene (**L**) is lethal in homozygous condition. Produces light coloured pigmentation in heterozygous condition (**L**l) and normal pigmentation in recessive condition (**ll**). What will be the phenotype for the F_1produced by a crossing of light coloured and normal pigmented fish?

Solution

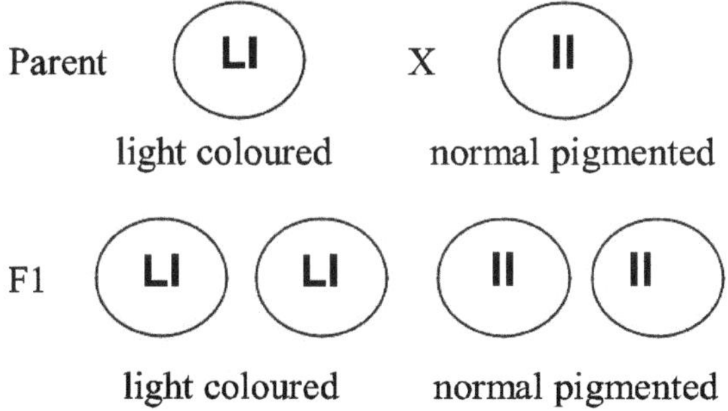

Answer

By crossing a light coloured and normal pigmented common carp the F_1 produced will be having light coloured and normal pigmentation in equal proportion.

Problem No. 3

The genotypes and phenotypes of different varieties of common carp are mentioned below:

SSnn, Ssnn - scaled or wild carp

ssnn - "scattered" mirror

SSNn, SsNn - linear mirror

ssNn - "leather"

SSNN, SsNN and ssNN are non-viable (lethal genes)

Calculate the phenotypic and genotypic ratio of the offsprings produced from the following crosses of common carp. Also mention the phenotype.

- ✰ ssnnxssNn
- ✰ ssNn x ssNn
- ✰ ssNn x SSnn
- ✰ ssnn x ssnn

Solution

a) ssnnxssNn

Parent ssnn x ssNn

Scattered mirror Leather

F1

	s,N	s,n
s,n	ssNn Leather	ssnn Scattered mirror
s,n	ssNn Leather	ssnn Scattered mirror

b) ssNn x ssNn

Parent ssNn x ssNn

Leather Leather

F1

	s,N	s,n
s,N	ssNN Lethal	ssNn Leather
s,n	ssNn Leather	ssnn Scattered mirror

c) ssNnx SSnn

Parent ssNnx SSnn

Linear mirror scaled

F1

	S,n	S,n
s,N	SsNn Linear mirror	SsNn Linear mirror
s,n	Ssnn Scaled	Ssnn Scaled

d) ssnn x ssnn

Parent ssnn x ssnn
Scattered
Mirror

F1

	s,n	s,n
s,n	ssnn Scattered Mirror	ssnn Scattered Mirror
s,n	ssnn Scattered Mirror	ssnn Scattered Mirror

Answers

a) Scattered mirror x leather
   Phenotypic ratio – leather: scattered mirror (1:1)
   Genotypic ratio – ssNn: ssnn (1:1)

b) Leather x leather
   Phenotypic ratio – 1 lethal: 2 leather: 1 scattered mirror
   Genotypic ratio – 1 ssNN: 2 ssNn: 1 ssnn

c) Leather x scaled
   Phenotypic ratio – linear mirror: scaled (1:1)
   Genotypic ratio – SsNn:Ssnn (1:1)

d) Scattered mirror x scattered mirror
   Phenotypic ratio – All scattered mirror Genotypic ratio – All ssnn

Chapter 4

Problems on Dihybrid Cross

Introduction

Dihybrid Cross

When a cross is made between two parents which are different in two pairs of characters such a cross is called as Dihybrid Cross.

Following the experiments with the monohybrid cross, Mendel tried to find out how different phenotypic traits behave in relation to each other in the inheritance from generation to generation. For this he crossed two varieties of pea plants differing in two pairs of contrasting characters. When round yellow plants were crossed with wrinkled green plants all the F_1 were Round Yellow. When F_1 were selfed the F_2 offspring produced were in the ratio of 9:3:3:1.

The irregularity of F_2 offspring ratio was explained by his second law,

Law of Independent Assortment or Recombination of Genes

It states that "when the gametes are formed the members of the different pairs of factors (alleles) for different traits segregate quite independently of each other and that all possible combination of the factors (alleles) concerned will be found among the progeny, as they are distributed to sex cells independently of one another"*E.g.* The allele **Y** was associated with allele **R** in parent but it does not always remain associated with it and, also becomes associated with allele **r**.

Mendel's genes which underwent independent assortment are located on different chromosomes which recombine freely in meiosis. But the segregation of genes that are present in same locus are governed by linkage or crossing over.

Significance of Mendel's Laws

- Mendel's most important discovery is his law of segregation. Segregation is the essence of Mendelism and a fundamental rule of genetics.
- The law of independent assortment holds good only for genes present in different homologous pairs of chromosomes.
- Inheritance of genes, present in the same homologous pair of chromosomes, is governed by the phenomenon of ***linkage** and **crossing-over***.
- Genes present in the sex chromosomes obey the rules of sex-linked inheritance.

Examples for Dihybrid Inheritance in Fishes

When two or more genes are inherited independently (*i.e.*, they are not linked), each phenotype is also inherited independently.

- For *e.g.* **gold** body coloration in the guppy is controlled by a simple autosomal recessive gene **g**. The dominant**G** allele produces **grey** guppies. **Curvature of the spine** is also controlled by a unique autosomal recessive gene **cu**. The dominant **Cu** allele produces **normal spines**.
- The two genes are inherited independently (law of independent assortment). The 9:3:3:1 phenotypic ratio is always produced when two heterozygotes mate and the two phenotypes are controlled by single autosomal genes with complete dominant gene action.
- In minnows the colour and finnage traits were independently controlled by two distinct loci, and within each locus the novel alleles, for pink colouration or long fins, were recessive to those producing the normal phenotype (grey colour and normal fin). Cross of a two heterozygous grey (GgNn x GgNn) normal results in the production of grey normal-9/16, grey long fin- 3/16, pink normal- 3/16 and pink long fin-1/16.

Interaction of Complementary Genes

The Interaction of Complementary genes occurs when two independent genes each Produce two separate Phenotypes, and the simultaneous occurrence of these traits Produces "new" or different Phenotypes that can be produced only by the simultaneous expression of the two traits. For example, body colour in fish is often produced by the simultaneous expression or non-simultaneous expression of different pigment cells, each of which is controlled by independent genes.

When this type of interaction occurs, the F2 Phenotypic ratio is the classic 9:3:3:1 ratio that is observed for the simultaneous expression of two independently segregating genes, each of which exhibits complete dominance and each of which controls a unique phenotype.

Epistatic Interaction

Epistasis is a type of gene interaction where an allele at one locus modifies or suppress the phenotypic expression of an allele at another locus. Epistatic interaction between two loci produces variations on the 9:3:3:1 F2 phenotypic ratio that occurs when there are two dominant genes which produce different phenotypes. When there is epistasis, the number of F2 phenotypes is reduced from four to either two or three, depending on the type of epistasis.

Body colour in many tropical fish is controlled by epistatic interactions between or among two or more loci. Body and fin colour in the Siamese fighting fish, examples of phenotypes that are controlled by epistatic interactions among four genes.

Dominant Epistasis

Dominant epistasis occurs when the dominant allele at one locus (the epistatic locus) produces a particular phenotype, regardless of the genotype at the second locus. In other words, the dominant allele at the epistatic locus prevents the second gene from producing either of its phenotype. The second gene can express its phenotypes only when the epistatic locus is homozygous recessive.

Problem No. 1

In goldfish, orange red colour (B) is dominant over blue (b) and normal eyes (D) over telescopic eyes (d). If the alleles segregate independently, write the most probable genotype for the parents for each of the following crosses by using punnett square method.

Sl.No.	*Phenotypes of Parents*	*Phenotypes of the Offspring (F2)*			
		ORN	*ORT*	*BN*	*BT*
1.	ORT x ORT	0	320	0	100
2.	ORN x ORT	180	190	0	0
3.	ORT x BN	200	0	210	0
4.	BN x BN	0	0	280	90

ORN: Orange red normal; BN: Blue normal; ORT: Orange red telescopic; BT: Blue Telescopic.

Solution

Approximate phenotypic ratio is calculated as follows:

Sl.No.	*Phenotypes of Parents*	*Phenotypes of the Offspring (F2)*			
		ORN	*ORT*	*BN*	*BT*
1.	ORT x ORT	0	3	0	1
2.	ORTx ORN	1	1	0	0
3.	BNx ORT	1	0	1	0
4.	BN x BN	0	0	3	1

Based on the phenotypic ratios calculated above, the most probable genotypes of the parents are calculated as follows:

1.

Parent

Bb dd x Bb dd

Orange red Telescopic Orange red Telescopic

F2

	B,d	B,d	b,d	b,d
B,d	BBdd ORT	BBdd ORT	Bbdd ORT	Bbdd ORT
B,d	BBdd ORT	BBdd ORT	Bbdd ORT	Bbdd ORT
b,d	Bbdd ORT	Bbdd ORT	bbdd BT	bbdd BT
b,d	Bbdd ORT	Bbdd ORT	bbdd BT	bbdd BT

2.

Parent Bbddx BB Dd

Orange red Telescopic Orange red Normal

F2

	B.D	B,d	B,D	B, d
B.d	BBDd ORN	BBdd ORT	BBDd ORN	BBdd ORT
B,d	BBDd ORN	BBdd ORT	BBDd ORN	BBdd ORT
b,d	BbDd ORN	BbDd ORT	BbDd ORN	BbDd ORT
b,d	BbDd ORN	Bbdd ORT	BbDd ORN	Bbdd ORT

3.

Parent bb DD x Bb dd

Blue Normal Orange Red Telescopic

F2

	B.d	B,d	b,d	b, d
b.D	BbDd ORN	BbDd ORN	bbDd BN	bbDd BN
b,D	BbDd ORN	BbDd ORN	bbDd BN	bbDd BN
b,D	BbDd ORN	BbDd ORN	bbDd BN	bbDd BN
b,D	BbDd ORN	BbDd ORN	bbDd BN	bbDd BN

4.

Parent bb Dd x bb Dd

Blue Normal Blue Normal

F2

	b.D	b,d	b,D	b, d
b.D	bbDD BN	bbDd BN	bbDD BN	bbDd BN
b,d	bbDd BN	bbdd BT	bbDD BN	bbdd BT
b,D	bbDD BN	bbDd BN	bbDD BN	bbDd BN
b,d	bbDd BN	bbdd BT	bbDd BN	bbdd BT

Answers

The most probable genotypes of the parents are as follows:

- ✰ BbddxBbdd
- ✰ Bbddx BBDd
- ✰ bbDD x Bb dd
- ✰ bbDdx bb Dd

Problem No. 2

Normal spine in guppy is governed by a dominant gene (Sn) and curved spine by its recessive allele (Sc). Grey colour results from the action of the dominant genotype (GG/Gg) and gold from the recessive genotype (gg). Determine the expected genotypic and phenotypic ratios in the progeny for the following crosses.

- ✰ SnSc Gg x SnSc Gg
- ✰ ScScGgxSnSc GG
- ✰ SnSnggxSnScgg
- ✰ ScSc gg x ScScgg

Solution

NSG: Normal spine Grey

NSGL: Normal Spine Gold

CSG: Curved spine Grey

CSGL: Curved spine Gold

- SnScGg x SnScGg

  NSG NSG

  F2

	Sn,G	Sn,g	Sc,G	Sc,g
Sn,G	SnSnGG NSG	SnSnGg NSG	SnScGG NSG	SnScGg NSG
Sn,g	SnSnGg NSG	SnSngg NSGL	SnScGg NSG	SnScgg NSGL
Sc,G	SnScGG NSG	SnScGg NSG	ScScGG CSG	ScScGg CSG
Sc,g	SnScGg NSG	SnScgg NSGL	ScScGg CSG	ScScgg CSGL

- ScScGgxSnSc GG

  NSG CSG

  F2

	Sn,G	Sn, G	Sc,G	Sc,G
Sc,G	SnScGG NSG	SnScGG NSG	ScScGG CSG	ScScGG CSG
Sc,g	SnScGg NSG	SnScGg NSG	ScScGg CSG	ScScGg CSG
Sc,G	SnScGG NSG	SnScGG NSG	ScScGG CSG	ScScGG CSG
Sc,g	SnScGg NSG	SnScGg NSG	ScScGg CSG	ScScGg CSG

- SnSnggxSnScgg

NSGL NSGL F2

	Sn,g	Sn, g	Sc,g	Sc,g
Sn,g	SnSngg NSGL	SnSngg NSGL	SnScgg NSGL	SnScgg NSGL
Sn,g	SnSngg NSGL	SnSngg NSGL	SnScgg NSGL	SnScgg NSGL
Sn,g	SnSngg NSGL	SnSngg NSGL	SnScgg NSGL	SnScgg NSGL
Sn,g	SnSngg NSGL	SnSngg NSGL	SnScgg NSGL	SnScgg NSGL

- ScScgg x ScScgg

CSGLCSGL

F2

	Sc,g	Sc, g	Sc,g	Sc,g
Sc,g	ScScgg CSGL	ScScgg CSGL	ScScgg CSGL	ScScgg CSGL
Sc,g	ScScgg CSGL	ScScgg CSGL	ScScgg CSGL	ScScgg CSGL
Sc,g	ScScgg CSGL	ScScgg CSGL	ScScgg CSGL	ScScgg CSGL
Sc,g	ScScgg CSGL	ScScgg CSGL	ScScgg CSGL	ScScgg CSGL

Answers

a) Phenotypic ratio:

Normal spine grey 9: Normal spine gold 3: Curved spine grey 3: Curved spine gold 1

Genotypic ratio:

SnSn GG (1):SnSnGg (2): SnSc GG (2): SnScGg (4): SnSngg (1): SnSc gg (2): ScSc GG (1): ScScGg (2): ScScgg (1)

b) Phenotypic ratio:

Normal spine Grey: Curved spine grey (1:1)

Genotypic ratio:

SnScGG: SnScGg: ScSc GG: ScScGg

1:1:1: 1

c) Phenotypic ratio:

All normal spine gold

Genotypic ratio:

SnSngg: SnScgg (1:1)

d) Phenotypic ratio:

All curved spine gold

Genotypic ratio:

All ScSc gg

Problem No. 3

In the case of minnows Grey-green **(G)** colour is dominant to pink colour. Normal fin (N) is dominant to long fin. If a homozygous Gray-green with normal fin male minnow is crossed with a homozygous pink and long fin female What will be the genotypes and phenotypes of the F_1 and F_2 offsprings.

Solution

GGN: Grey green normal

GGL: Grey green long fin

PN: Grey green normal

PL: Pink long fin

Parent GGNN x ggnn

GGN PL

F1 GgNnGgNnGgNnGgNn

All Grey Green Normal

F2

	G,N	G,n	g,N	g,n
G,N	GGNN GGN	GGNn GGN	GgNN GGN	GgNn GGN
G,n	GGNn GGN	GGnn GGL	GgNn GGN	Ggnn GGL
g,N	GgNN GGN	GgNn GGN	ggNN PN	ggNn PN
g,n	GgNn GGN	Ggnn GGL	ggNn PN	ggnn PL

Answers

- ☆ F_1 Phenotype:
  All Grey-Green Normal fin

  Genotype:
  GgNn

- ☆ F_2 Phenotype:
  Grey Normal fin -9/16,
  Grey long fin -3/16,
  Pink normal -3/16,
  Pink long fin -1/16

Genotype:

☆ NN (1): GG Nn (2): Gg NN (2): Gg Nn (4): GG nn (1): Ggnn (2): ggNN (1): ggNn (2): ggnn (1)

Problem No. 4

In the case of Nile tilapia, two loci **A** and **B** are involved in producing normal body colouration. The recessive alleles **a** and **b** in combination, and only in combination, would result in pearl colouration. Either recessive allele, when present alone, would not alter the normal colouration of the fish. Determine the expected phenotypic ratios in the progeny from the following mating.

☆ aa bbxAA BB

☆ Aa Bb x Aa Bb

☆ AA BbxAa BB

☆ aa Bb x aa Bb

Solution

(i) aa bbxAA BB

Normal Pearl

F2

	A,B	A,B	A,B	A,B
a,b	AaBb Pearl	AaBb Pearl	AaBb Pearl	AaBb Pearl
a,b	AaBb Pearl	AaBb Pearl	AaBb Pearl	AaBb Pearl
a,b	AaBb Pearl	AaBb Pearl	AaBb Pearl	AaBb Pearl
a,b	AaBb Pearl	AaBb Pearl	AaBb Pearl	AaBb Pearl

a] Aa Bb x Aa Bb

Pearl Pearl

F2

	A,B	A,b	a,B	a,b
A,B	AA BB Normal	AA Bb Normal	AaBB Normal	AaBb Pearl
A,b	AA Bb Normal	AA bb Normal	AaBb Pearl	Aa bb Pearl
a,B	AaBB Normal	AaBb Pearl	aaBB Normal	aaBb Pearl
a,b	AaBb Pearl	Aa bb Pearl	aaBb Pearl	aa bb Pearl

c) AA BbxAa BB

Normal Normal

F2

	A,B	A,B	a,B	a,B
A,B	AA BB Normal	AA BB Normal	AaBB Normal	AaBB Normal
A,b	AA Bb Normal	AA Bb Normal	AaBb Pearl	AaBb Pearl
A,B	AA BB Normal	AA BB Normal	Aa BB Normal	Aa BB Normal
A,b	AA Bb Normal	AA Bb Normal	Aa Bb Pearl	Aa Bb Pearl

mm aa Bb x

aa Bb Normal

Normal F2

	a,B	a,b	a,B	a,b
a,B	aaBB Normal	aaBb Pearl	aaBB Normal	aaBb Pearl
a,b	aaBb Pearl	aabb Pearl	aaBb Pearl	aabb Pearl
a,B	aaBB Normal	aaBb Pearl	aaBB Normal	aaBb Pearl
a,b	aaBb Pearl	aabb Pearl	aaBb Pearl	aa bb Pearl

Answers

The expected phenotypic ratios are as follows:

- ☆ All pearl
- ☆ Normal-7/16, Pearl -9/16

  Normal-3/4, Pearl -1/4

  Pearl -3/4, Normal-1/4

Problem No. 5

In gold fish, the gene **M** is epistatic with respect to the gene **S**. Genes **m** and **s** produce albino.

The following combinations of genes are possible.

MS and Ms-dark fishes,

mS-light fishes,

ms-albino fishes.

By using the punnet square find out the phenotypic and genotypic ratio for the mating of the gold fish with following genotypes.

- ☆ Two heterozygous MmSs dark gold fish
- ☆ mmss x Mm Ss
- ☆ mmss x MM Ss
- ☆ MM Ss x Mm Ss
- ☆ MM ss x Mm Ss

Solution

MS and **Ms** - dark fishes,

mS - light fishes,

ms-albino fishes.

a) Two heterozygous MmSs dark gold fish

	M S	M s	m S	m s
M S	MMSS dark	MMSs dark	MmSS dark	MmSs dark
M s	MMSs dark	MMss dark	MmSs dark	Mmss dark
m S	MmSS dark	MmSs dark	mmSS light	mmSs light
m s	MmSs dark	Mmss dark	mmSs light	mmss dark

b) mmssxMm Ss

Albino dark

	M S	M s	m S	m s
m s	MmSs dark	Mmss dark	mmSs light	mmss albino
m s	MmSs dark	Mmss dark	mmSs light	mmss albino
m s	MmSs dark	Mmss dark	mmSs light	mmss albino
m s	MmSs dark	Mmss dark	mmSs light	Mmss albino

c) mmssxMM Ss

Albino dark

	M S	M s	M S	M s
m s	MmSs dark	Mmss dark	MmSs dark	Mmss dark
m s	MmSs dark	Mmss dark	MmSs dark	Mmss dark
m s	MmSs dark	Mmss dark	MmSs dark	Mmss dark
m s	MmSs dark	Mmss dark	MmSs dark	Mmss dark

d) MM Ss x Mm Ss

darkdark

	M S	M s	M S	M s
MS	MMSS dark	MMSs dark	MMSS dark	MMSs dark
Ms	MMSs dark	MMss dark	MMSs dark	MMss dark
mS	MmSS dark	MmSs dark	MmSS dark	MmSs dark
ms	MmSs dark	Mmss dark	MmSs dark	Mmss dark

e) MM ssxMm Ss

darkdark

	MS	Ms	mS	Ms
M s	MMSs dark	MMss dark	MmSs dark	Mmss dark
M s	MMSs dark	MMss dark	MmSs dark	Mmss dark
M s	MMSs dark	MMss dark	MmSs dark	Mmss dark
M s	MMSs dark	MMss dark	MmSs dark	Mmss dark

Answer

a) Phenotypic ratio: 12 dark: 3 light: 1 albino
Genotypic ratio: 12 MS or Ms: 3 mS: 1ms

b) Phenotypic ratio: 2 dark:1light:1 albino
Genotypic ratio: 2 MS or Ms:1mS:1ms

c) Phenotypic ratio: All dark
Genotypic ratio: all MS or Ms

d) Phenotypic ratio: All dark
Genotypic ratio: all MS or Ms

e) Phenotypic ratio: All dark
Genotypic ratio: all MS or Ms

Problem No. 6

The phenotype and genotype of different varieties of platy fish are given as below:

Genotype	*Phenotype*
StSt, RR	normally pigmented
StSt, Rr	normally pigmented
StSt, rr	gray
Stst, RR	normally pigmented

Genotype	*Phenotype*
Stst,Rr	normally pigmented
Stst, rr	gray
stst, RR	gold
stst, Rr	gold
stst, rr	brown

Find out the phenotypic ratio of the F2 progeny for the cross between two heterozygous (Stst, Rr) normally pigmented platy fish.

Solution

F2

	St, R	St, r	st, R	st, r
St, R	StSt, RR normally pigmented	StSt, Rr normally pigmented	Stst, RR normally pigmented	Stst, Rr normally pigmented
St, r	StSt, Rr normally pigmented	StSt, rr gray	Stst, Rr normally pigmented	Stst, rr gray
st, R	Stst, RR normally pigmented	Stst, Rr normally pigmented	stst, RR gold	stst, Rr Gold
st, r	Stst, Rr normally pigmented	Stst, rr gray	stst, Rr gold	stst,rr brown

Answers

The phenotypic ratio of F2 progeny resulted from the cross of two heterozygous (Stst, Rr) normally pigmented platy fish is 9:3:3:1 (normally pigmented: gray: gold: brown)

Chapter 5

Induction of Triploidy in Fishes

Introduction

Triploidy is a process in which there will be addition of one set of chromosomes to the original diploid complement by preventing the extrusion of second polar body of the fertilized egg. Triploidy is induced by subjecting the fertilized egg by normal sperm to the usual shock treatments. Generally triploids are supposed to be sterile. Sterility is great advantage in modern fish culture, as energy spent for maturation or gonadal development may possibly diverted or utilized for increased somatic growth, especially species like common carp and tilapia, which have shorter maturity cycle. Triploid common carp have shown more than 60 per cent growth rate than its normal diploid counterpart. Because of its sterility overpopulation is also checked, as the species has pond breeding habits similar to tilapias. Sterile triploid grass carp can be safely used for controlling aquatic weeds in open water system without fear of its establishment through reproduction.

Natural Triploidy

Polyploidy has been observed to occur in nature in some species of fish like the common carp (*Cyprinus carpio*) and trout mainly due to chromosomal translocation and when two distantly related fish species are crossed. The crosses between grass carp and bighead carp had produced triploid hybrids (Marian and Krasznai, 1978). However, none of the interspecific nor intergeneric hybrid crosses among Indian carps have been reported to produce such allotriploids.

Artificial Induction of Triploidy in Indian Major Carps

Reddy *et al.* (1987), made the first attempt of inducing polyploidy in one of the Indian major carps, *Labeo rohita* by using colchicine. They could induce only tetraploid and diploid mosaics. However, they successfully induced triploidy in rohu, *L.rohita*, and tetraploidy in *L. rohita* and *Catla catla* by using thermal shocks. Similarly successful induction of tetraploidy was also reported in the case of *Cirrhinus mrigala* and *L.rohita* (Zhang, 1990). Triploidy was induced in rohu by administering heat shocks to zygotes, seven minutes after fertilization at 42C +0.5 C for a duration of 1–2 minutes. However, the incidence of triploidy was found to be around 12 per cent. Most of the eggs died probably due to lethal temperature level. They also reported that rohu zygotes exposed to heat shock, prior to first cleavage at 39C+0.5C for two minutes yielded 70 per cent tetraploids and those exposed to cold shocks of 10–15C for 10 min. yielded only 30–55 per cent tetraploids. In the case of catla, heat shocks to zygotes at 4°C exposed for two minutes prior to first cleavage yielded 30–65 per cent tetraploids, but cold shocks were not effective.

Zhang in the year 1990 induced tetraploidy in *C.mrigala* for the first time and also in *L.rohita*. He reported that heat shocks administered at 39–4°C for two minutes in embroys 22–25 min post-fertilization could induce tetraploidy in mrigal and rohu. However, the percentage of polyploids ranged from only 10–40. Induction of triploidy in *L. rohita* by applying heat shock at 4°C for 2 min starting 4 min after fertilization was also reported. The rate of triploidy induction was found to be 60 per cent. The survival in the heat shock groups was recorded 15 per cent and 25 per cent in the control groups. The growth rate of triploid individuals was found to be significantly higher than in diploids for weight and length, $P<0.01$ and $P<0.05$ respectively.

Induction of Triploidy

Mechanism for triploid induction occurs after the eggs are ovulated and the sperm enters the eggs. Before ovulation, the nuclei of eggs in the ovary of fish have 4n number of chromosomes. Ovulation starts with the disappearance of the nuclear membrane in the egg and the appearance of chromosomes, leading to the first meiotic division. The first polar body is formed and pushed out of the egg, resulting in a reduction of chromosomes to 2n. The follicles, which attach the eggs within the ovary, split and partially dissolve, releasing the eggs. The eggs then flow through the genital opening of the female fish. After the sperm enters the egg through the micropyle, the genetic content of the nucleus of the egg is further reduced to 1n when the second polar body is formed (second meiotic division) and ejected from the egg. The genetic material of the sperm (in) and the nucleus of the egg (In) then combine to form a developing embryo. Triploid fish have been produced by preventing the second meiotic division of the egg (after the sperm enters the egg). Therefore, two sets of chromosomes are contributed by the female and one set + by the male (2n egg in sperm = 3n). This procedure is usually accomplished through chemical, thermal, or mechanical methods. Chemical methods are primarily experimental and restricted to research laboratories. Thermal and mechanical methods are commonly used in production facilities. Triploid fish are commercially produced by applying thermal or pressure shocks shortly after water is added to

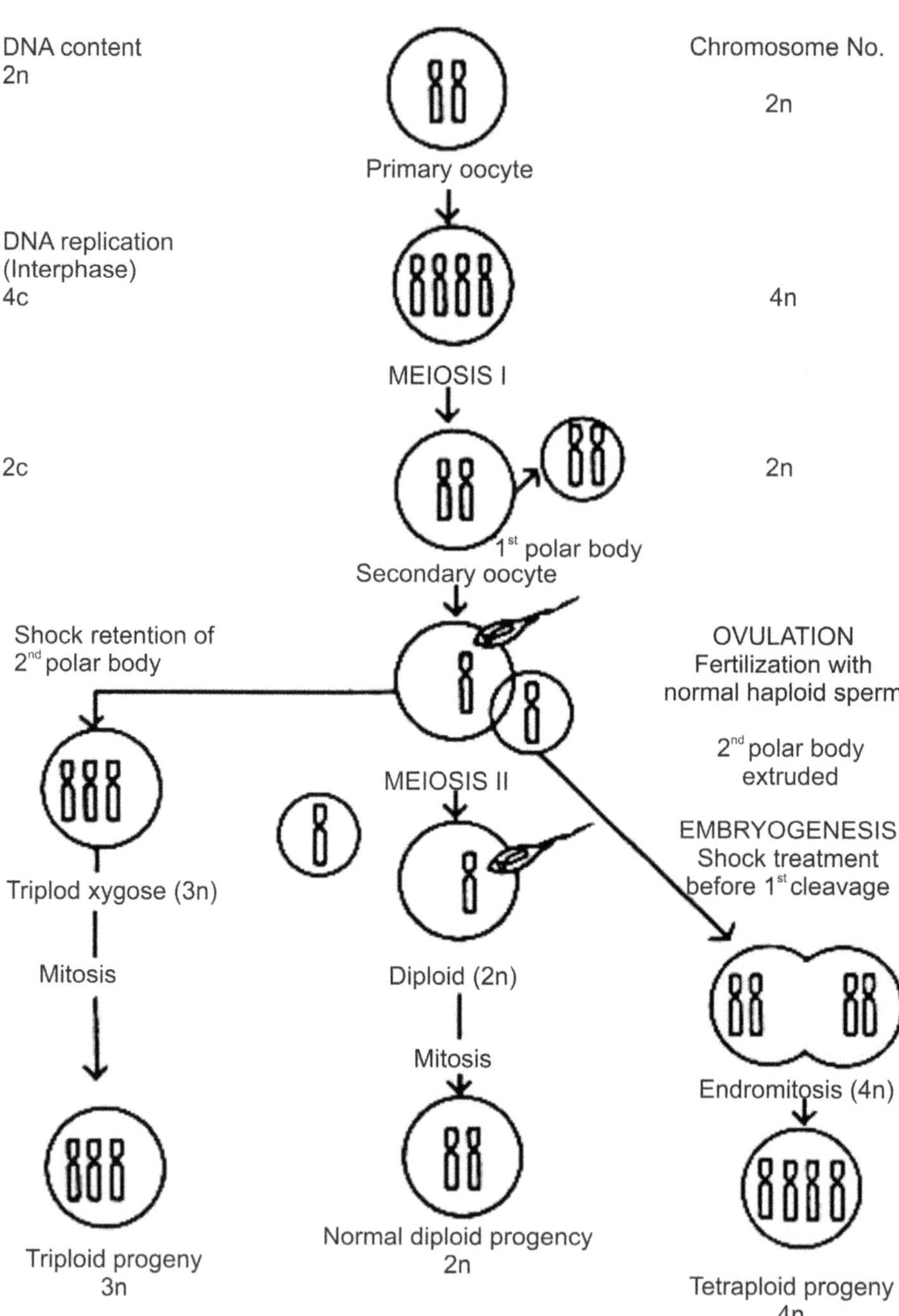

Induction of Triploidy.

the egg and sperm mixture. The precise timing of the second meiotic division and thus, the application of treatment to induce triploidy, varies with the species of fish and water temperature

Verification of Triploidy

Because hydrostatic pressure treatments seldom result in 100 per cent triploid individual fish, each individual must be verified triploid before they can be stocked in the waters of many states. Triploidy in grass carp is commonly verified by taking a blood sample and analyzing the volume of red blood cells, using an electronic particle counter. Blood samples can be taken from grass carp greater than 50 millimeters (2 inches) long. Fish may be anesthetized if necessary. The head of the fish is tilted back and the branchial artery on the isthmus is punctured with a blood lancet rinsed in EDTA between samples. Approximately 1 microliter of blood is withdrawn using either a positive displacement micropipette with disposable polypropylene tips or disposable hematocrit capillary tubes. The blood sample is immediately placed in a vial containing an electrolyte solution (*e.g.*, HematallAzide-Free Isotonic Diluent™, Fisher Scientific®). The cell membrane of the red blood cells are dissolved (lysed)with a lysing agent (*e.g.*, Hematall LA-Hgb Reagent™, Fisher Scientific®), leaving the nuclei. The lysed blood cells are immediately scanned for ploidy determination with an electronic particle counter, calibrated to read both diploid and triploid red blood cell nuclei volumes. Diploid grass carp have a red blood cell nuclei volume of 10.06 cubic micrometers (μm^3),while the mean volume of triploid red blood cell nuclei is 14.82 μm^3.

Materials Required

Matured male and female fishes, fish breeding hormones, breeding and hatchery accessories, cotton, small screw cap tubes, petridish, ice blocks, water-bath, thermometer, quill feather, Hank's Balanced Salt Solution, *etc.*

Procedure

1. Inject spawning hormone such as ovaprim/ovatide/WOVA FH to the brood fishes in the late evening hours.
2. Strip the milt from ripened males after allowing sufficient latency period (12 hrs.)
3. Mix milt with freshly prepared Hank's solution 1:4 without bicarbonate and store at 4°C for 1-6 hrs. or until needed.
4. Strip the matured eggs as soon as ovulation commences.
5. Mix the stored milt (before mixing, the milt should be brought to water temperature) with freshly stripped eggs and fertilize the eggs by dry fertilization method.
6. Treat the fertilized eggs to cold shock (0°C for 1 hr.) or heat shock (40°C for 1 or 2 min) immediately after fertilization before the release of second polar body.

7. After shock treatment bring the treated eggs to normal condition and incubate them for hatching.
8. Estimate the percentage of triploidy induction by preparing the chromosome spreads or by calculating the nuclear and cellular area and volume.

Chapter 6

Induction of Gynogenesis in Fishes

Introduction

Gynogenesis is the process of embryonic development with solely the maternal genome and without paternal genetic input, a phenomenon similar to parthenogenesis. Gynogenesis occurs in nature and can be also induced.Gynogenesis is the process of embryonic development with solely maternal (female parent) genome and without paternal (male parent) genetic input. The main objective of gynogenesis is to produce highly homozygous inbred lines in much shorter time than through the conventional breeding process. These gynogen lines can be top-crossed with the heterozygous stock to achieve heterosis effect. Gynogenesis can be artificially induced by eliminating/denaturing the genetic material (DNA) of the milt through irradiation either by UV or gamma rays and activating the eggs with irradiated milt. Diploidy is restored by giving thermal or hydrostatic pressure shock. Retention of diploidy in a haploid gynogen egg can be done by following two ways.

1. By suppressing metaphase II in the second meiotic division thereby preventing the extrusion of II Polar body (meiotic gynogenesis).
2. By blocking the 1st mitotic cleavage (mitotic gynogenesis)

The degree of homozygosity in meiotic gynogens is supposed to be 50 per cent and mitotic gynogens is 100 per cent. For inducing meiotic gynogenesis shock treatment should be given immediately after fertilization (5 min). For mitotic gynogenesis the shock treatment can be given around 20-30 min after fertilization.

Natural Gynogenesis

Natural gynogenesis has been reported in some members of the family Poecilidae, *e.g. Poecilia formosa* and *Cyprinidae e.g. Carassius auratus gibelio* (Gold,

1979). The cross between certain species of *Cyprinidae* such as the Chinese grass carp and silver carp also resulted in gynogenetic offsprings.

Artificial (Induced) Gynogenesis

Artificial induction of gynogenesis involves genetic inactivation or destruction of DNA content of milt by exposing it to either ultraviolet or gamma rays and activation of eggs with this genetically inactivated milt. However, these activated eggs develop into haploid embryos/hatchlings and do not survive unless diploidy is restored by subjecting the activated eggs to either thermal or pressure shocks at an appropriate stage after activation, just before the extrusion of the polar body (meiotic gynogenesis) or blocking the first mitotic division in the developing/ activated egg (mitotic gynogenesis). The pace of development may differ from species to species and depends on the ambient temperature. Artificial gynogenesis has been successfully induced in many species of fish including carps.

Induction of Meiotic Gynogenesis in Indian Major Carps

Meiotic gynogenesis was successfully induced in Indian major carps for the first time in *L. rohita* and *C. catla* in the year 1981 (John *et al.,* 1984) and in *Cirrhinus mrigala* (. The procedure includes, genetic or DNA denaturalization or inactivation in the penetrating spermatozoa by exposure to either ultraviolet or gamma rays. For inactivating genetic component of the male gametes, ultraviolet irradiation is a simple device and very effective in fishes. To restore diploidy in the activated egg, either thermal stocks (heat or cold) or pressure shock are usually administered. However, the efficacy of these shocks, especially thermal shock, and their intensity and duration of treatment seems to vary from species to species. The effective intensity lies at some point around the sub-lethal levels. Thus in all the species of Indian carps, cold shocks below 10°C and heat shocks above 42°C proved to be lethal (John *et al.,* 1984 and 1988 and Reddy *et al.* 1993). Effective genetic inactivation of sperm DNA was possible in Indian major carps by exposing the milt, diluted to 1:4 with cold Hank's solution, (milt 1 and Hank's 4 parts), to a 15w ultraviolet germicidal tube at a distance of 20 cm and irradiated for 17–20 min (John *et al.,* 1984). The milt was poured into petri dishes fixed on ice blocks, keeping the column depth of milt to 2mm and placing the petri dish on a mechanical shaker for irradiation. It is preferable to use milt from a different species since improper irradiation can be detected by the presence of hybrids. In the case of Indian major carps meiotic gynogenes of rohu were produced by activating the eggs with the irradiated milt of catla and vice versa (John *et al.,* 1984) and mrigal gynogens were produced by activating the eggs with the irradiated milt of common carp (John *et al.,* 1988).

The effective intensities of cold and heat shocks for inducing diploid meiotic gynogenesis in Indian major carps were 12°C for 10 min and 39°C for one to two minutes respectively. The appropriate time after activation for the administration of shock treatments to induce meiotic gynogenesis (retention of polar body) in the

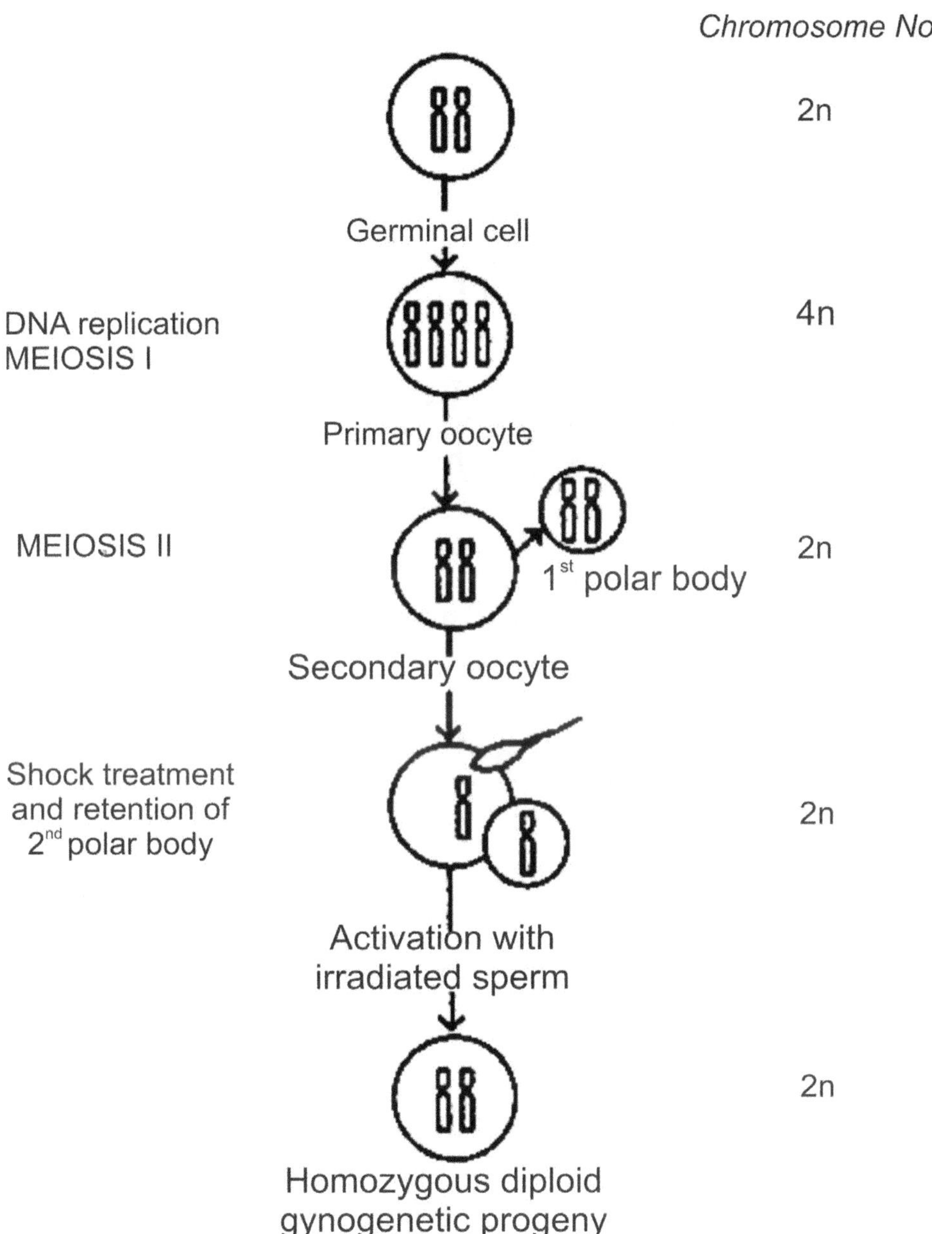

Shock Treatment and Retention of 2nd Polar Body (Meiotic Gynogenesis).

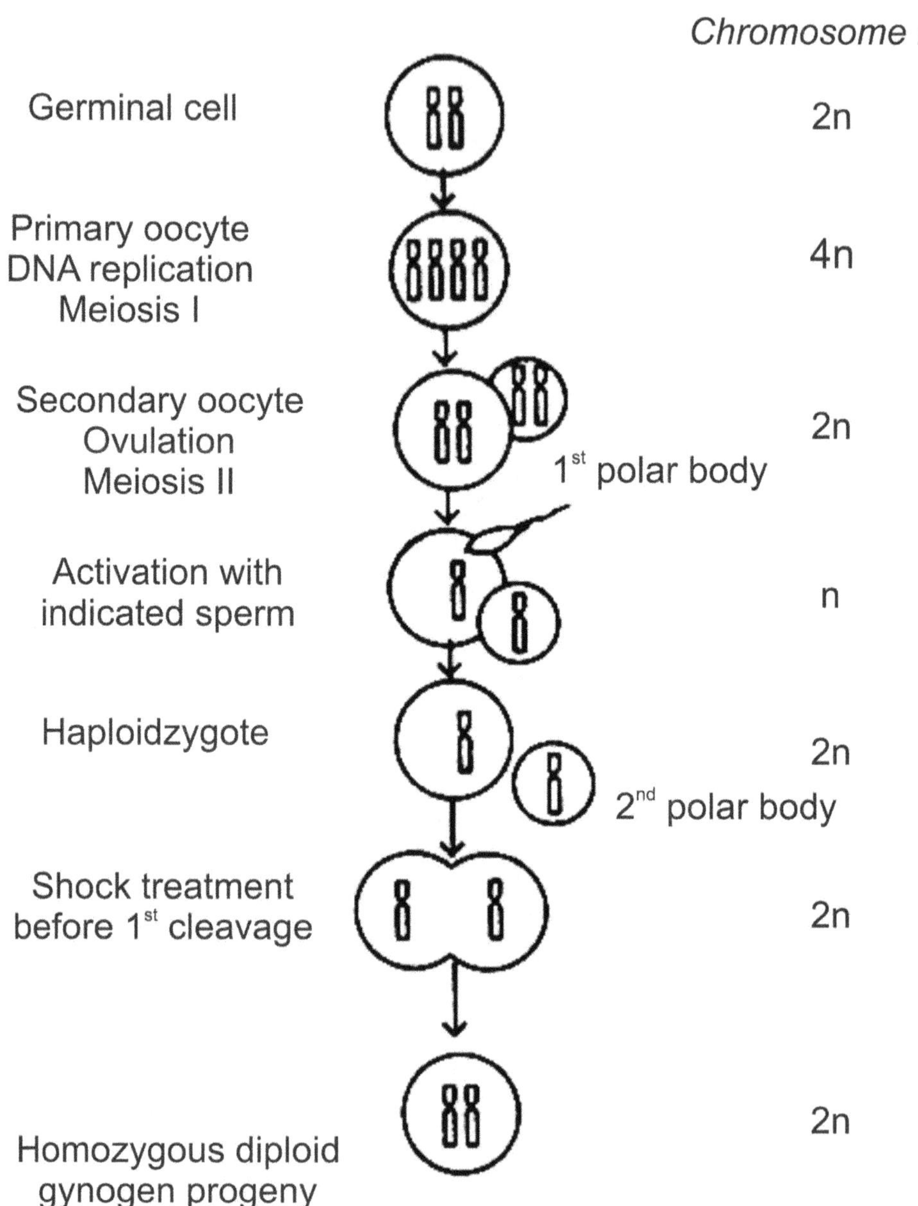

Shock Treatment before First Cleavage (Mitotic Gynogenesis).

activated egg of Indian carps was reported to be 4min after activation. Heat-shock treatment is shown in Plate 8. The percentage yield of meiotic gynogens ranged from 12.6 to 18.1 per cent (John *et al.,* 1984). In catla, heat-shocks were more effective than cold-shocks while the reverse was true in the case of rohu (John *et al.,* 1984).

Induction of Mitotic Gynogenesis in Indian Major Carps

The first report on successful induction of mitotic gynogenesis in rohu was given by Reddy *et al.,* 1993, followed by Hussai, *et al.,* 1994. They used irradiated milt of common carp, *Catla catla* and also *Cirrhinus mrigala*, to activate eggs of rohu (*Labeo rohita*) to produce mitotic gynogen rohu. Irradiated milt of Chinese grass carp (*Ctenopharyngodon idella*) was also used to induce mitotic gynogenesis in rohu, catla, mrigal and kalbasu. Use of milt from heterologous species, preferably having low compatibility to hybridize may be more desirable for the easy elimination of paternal genetic input with irradiation. Therefore, the milt from Chinese carps was used, as they are incompatible to produce viable hybrids with Indian carps.

While inducing mitotic gynogenesis in rohu, Reddy *et al.* (1993) followed the same cold shock regime as did John *et al.,* (1984) for meiotic gynogenesis. However, the heat shock regime was raised to 40°C keeping the duration of exposure same in both the shock treatments. Mitotic gynogen rohu, diploid and haploid hatchlings are shown in Plates 9 and 10 respectively.

Materials Required

Matured male and female fishes (Indian Major carps/catfishes), fish breeding hormones, breeding and hatchery accessories, cotton, ice-blocks, UV irradiation chamber, water bath, thermometer and microscope.

Procedure

- ☆ Inject the required spawning hormones to the brood fishes.
- ☆ After incubation period strip the milt freshly from male fish (preferably from another species) an hour before the expected ovulation time of the female fish (the advantage of using milt from different species is that it will be easier to distinguish the gynogens from that of hybrid. The eggs which are fertilized by an unaffected sperm by UV irradiation will develop into a hybrid).
- ☆ Mix the milt with freshly prepared Hank's solution. Divide the milt into two parts.
- ☆ Keep one half as control and other half for irradiation in a petridish to a 1-2 mm thick.
- ☆ Irradiate stored or freshly collected milt with UV-rays (254 nm, power 1.0 mw/cm at the surface of the milt) under a germicidal lamp for 15-20 min.
- ☆ Check the motility of the sperm after irradiation for different period.

- After irradiation store the milt in cold temperature or in ice.
- Strip the matured eggs as soon as the ovulation commences.
- Fertilize the freshly stripped eggs with normal and UV irradiated sperm separately.
- Subject the fertilized eggs to heat, cold shock to inhibit the release of II Polar body or 1st mitotic cleavage.
- Incubate the fertilized eggs in the hatchery.
- Rear the hatched larvae and analyse the chromosome number. Success of gynogenesis may be observed based no hybrid fish produced.
- Calculate the relative survival rates of the gynogenetic treatment groups and control groups (Survival of gynogenetic group/survival of control group). Monitor the survival at the eye stage, hatching, initiation of feeding and maturation.

Chapter 7

Preparation of Metaphase Chromosomes from Fish Tissues

Introduction

The study of fish chromosomes has been in wide use in the recent past mainly to ascertain the taxonomic position of different species. The manipulation of chromosomes has proved a valuable technique and its application in fish breeding and culture is yet to be used extensively. The study of chromosomes is also helpful for genome manipulation and banding studies. The karyotype of a normal individuals usually consists of two sets (diploid) of chromosomes derived one set from each parent. In the case of diploid gynogens, the karyotype consists of duplicated set of chromosomes derived by the retention of second polar body while all the three sets of chromosomes, *viz.*, the egg, the second polar body and the sperm together form the karyotype of a triploid individual.

Principle

The basic principle of getting metaphase chromosomes is to arrest a mitotically dividing cell at the subsequent stage. Here mitosis is induced in some heamopoetic tissues: *viz.*, kidney, spleen *etc.*, in the living system of a fish with a mitogen like concanavaline A (or may be with phytohaemagglutinin). This induced mitosis is allowed to continue *in vivo* for some specific time interval and then the cells are treated with colchicine. Colchicine is a potent inhibitor of tubulin polymerization and it dismantles the microtubule spindles. Thus the separation of the chromosomes is stopped. But the concentration and time of this treatment must be very critically maintained. Otherwise the chromosomes will be excessively contracted.

The tissues are isolated from the system after the colchicine treatment and minced to get a single cell suspension in hypotonic solution which also helps to swell the cell very much. They are fixed with Carnoy's solution (3:1 methanol, acetic acid) which helps to clear up the cytoplasmic content of the cells and the nucleolar proteins. Then the fixed cells are burst open on a cleaned/warmed slide by falling a drop of cell suspension from sufficient height (1-2 feet is sufficient).

Chemicals Required

1. Calcium chloride. Treatment with calcium chloride is believed to enhance mitotic division and to counteract the contraction of chromosomes normally produced by colchicine
2. Colchicine solution (0.05 per cent): 0.05g/100ml sterile double distilled water, @ 0.1 ml/10gm of body weight
3. Carnoy's solution - Methanol: Acetic acid (3:1)
4. Hypotonic solution; KCl: 0.56 per cent (0.075 M) in sterile distilled water or 0.9 per cent aqueous sodium citrate
5. Ethylene glycol – 50 mg/ml. to avoid animal suffering anesthetized.
6. Giemsa stain -Stock solution – Dissolve Giemsa powder in 33 ml glycerol and incubate overnight at 60°C in water bath, then cool it to room temperature and add 33 ml of methanol to the above solution. Filter and use as stock solution. Store the stain in amber coloured bottle at room temperature.

Working Solution

5 ml stock in 95 ml Phosphate buffer in 5 per cent Giemsa solution.

Phosphate Buffer Stock Solution

- ☆ 0.5M $KH_2 PO_4$ - 6.8g/100 ml d.H_2O
- ☆ 0.5M $Na_2 HPO_4$ - 17.9g/100 ml d.H_2O

Working Solution

- ☆ 0.5M $KH_2 PO_4$ - 6.26 ml
- ☆ 0.5M $Na_2 HPO_4$ - 4.56 ml

Dilute up to 500 ml with distilled water.

Equipment's Required

Cleaned glass slides, Cleaned glass centrifuge tubes (10 ml), Complete set of dissection equipment's all cleaned, Pasteur pipettes with bulbs, Cleaned cavity blocks one each for different samples of fishes, Parafilm strips, tissue paper, Nikon microscope, Tissue homogenizer and Spirit lamp.

Procedure

- ☆ Take a healthy fry or fish and condition them. Inject 0.1 per cent aqueous $CaCl_2$ solution through intraperitoneal injection (for 5-10 cm fish — 0.5ml)
- ☆ Then the fishes are to be injected intramuscularly with 0.01-0.05 per cent colchicine at a dosage of 0.5 ml to 1.0 ml/100 g body weight of the fish. If the fishes are small fry, add required quantity (0.1 per cent to 0.001 per cent) of colchicine in the aquarium.
- ☆ The fishes should be kept in well aerated aquaria for 1-2 hr after colchicine treatment.
- ☆ Dissect the fish and isolate the kidney (or/and spleen) for larger fish and gill arches for smaller fish by avoiding the upper membrane and blood clots.
- ☆ Immediately put the tissue in 6-8 ml hypotonic solution (0.56 per cent KCI or 0.08 per cent sodium citrate in sterile water or distilled water) in 15 ml centrifuge tubes.
- ☆ By using tissue homogenizer make the tissues to cell suspension (for hatchlings and early fry whole body tissues can be taken).
- ☆ Carefully decant the upper liquid cell suspension into a 15 ml clinical centrifuge tube and try to avoid the particulate piece which comes in contact. If any such piece of tissue comes then remove it from the cell suspension by carefully decanting into another tube. Gently with Pasteur pipette, mix it. Incubate it for 20-25 min at room temperature for swelling.
- ☆ Gently add 1.0 ml freshly prepared chilled Carnoy's fixative through the sides of the tube. Mix it gently with Pasteur pipette.
- ☆ Spin at 1500 rpm in clinical centrifuge for 10 min at room temperature with proper balancing to pellet down the cells from the suspension.
- ☆ Drain the supernatant with a pipette and slowly overlay 6-8 ml freshly prepared chilled fixative.
- ☆ Keep the tubes in refrigerator (at 4°C) for half an hour for thorough fixation (the duration of fixation may be prolonged further).
- ☆ Mix the contents and centrifuge cell suspension at 1200-1500 rpm for 10 min at room temperature. Remove the supernatant without disturbing the cell pellet at the bottom and add the fresh fixative (5-6 ml).
- ☆ Repeat the above step for 3-4 times till a clear transparent cell suspension is obtained.
- ☆ Take a clean glass slides (treated with 50 per cent ethanol). A drop of the cell suspension with fixative can be dropped from a Pasteur pipette held at a height of 1-1.5 feet to the slide.
- ☆ Allow the slide to air or flame to dry.

- Keep the slide for ageing for 1-3 days in dust free place.
- Stain the slide in 5 per cent Giemsa solution (pH 6.8) for 20 min by keeping the slides back to back into the staining jar.
- Wash the slide with distilled water thoroughly and store the slides in a slide box. Make the slides permanent by mounting in DPX mounting.
- Observe the slide under Nikon microscope under oil immersion objective (100X).

Chapter 8

Protocol for Chromosome Preparation from Embryos

Introduction

The study of fish chromosomes has been in wide use in the recent past mainly to ascertain the taxonomic position of different species. The chromosome preparation from embryos will help us to ascertain the proper structure and number of chromosome in the embryonic stage itself.

Chemicals Required

- ☆ Colchicine solution (0.05 per cent): 0.05g/100ml sterile double distilled water, @ 0.1 ml/10gm of body weight
- ☆ 0.7 per cent NaCl solution: 0.7 gram of NaCl in 100 ml of sterile double distilled water
- ☆ Ethanol: Acetic acid (4:1)
- ☆ Giemsa stain -Stock solution:

Dissolve Giemsa powder in 33 ml glycerol and incubate overnight at 60°C in water bath, then cool it to room temperature and add 33 ml of methanol to the above solution. Filter and use as stock solution. Store the stain in amber coloured bottle at room temperature.

Working Solution

5 ml stock in 95 ml Phosphate buffer is 5 per cent Giemsa solution.

Equipment's Required

Cleaned glass slides, Cleaned glass centrifuge tubes (10 ml), Complete set of dissection equipment's all cleaned, Pasteur pipettes with bulbs, Cleaned cavity blocks one each for different samples of fishes, Para film strips, tissue paper, Nikon microscope and Tissue homogenizer.

Procedure

- Take 1ml of colchicine and add 20 ml of hatchery water.
- Add the required number (3-4) of embryos to the colchicine solution and leave it for a minimum period of 4 hrs.
- Drain off the colchicine and add 0.7 per cent Na Cl.
- Remove the yolk sac from the embryos and put into distilled water for 20 min. This will swell the embryonic cells.
- Transfer the embryos into an ethanol/acetic acid 4; 1 mixture. They can be stored for up to 30 days at 4°C.
- Mince the embryos for 1 min with tissue homogenizer to dissociate epithelial cells.
- Drop the solution into the pre-warmed slide (45°C).
- Leave for 10 sec.
- Stain the slide in 5 per cent Giemsa stain for 30 min and wash the slide in tap water.
- Dry the slide and observe under the light microscope.

Chapter 9

Cryopreservation of Fish Sperm

Introduction

Cryopreservation is a long-term storage technique with very low temperatures to preserve the structurally intact living cells and tissues for extended period of time at a relatively low cost. Cryopreservation is to preserve and store the viable biological samples in a frozen state over a extended period of time. A very important part is that the research in cryopreservation is to reveal the underlying physical and biological responses of the cell and cause of cryoinjury, especially those associated with the phase change of water in extracellular and intracellular environments (Mazur 1984). From the original slow-cooling study, another cryopreservation approach has moved to easier and more efficient technique-vitrification, Cryoprotective agents has to gain access to all the parts of the system. Cryopreservation considers the effects of freezing and thawing. Therefore, the diffusion and osmosis processes have important effects during the introduction of cryoprotective agents, the addition or removal of cryoprotectants, the cooling process, and during thawing. These phenomena are amenable to the experimental design and analysis. Thus, reliable methods can be developed for preserving a very wide range of cells and some tissues. These methods have found to have widespread applications in biology, biomedical technology and conservation.

Germplasm cryopreservation includes storage of the sperm, eggs and embryos and contributes directly to animal breeding programmes. Germplasm cryopreservation also assist the *ex situ* conservation for preserving the genomes of threatened and endangered species. The establishment of germplasm banks using cryopreservation can contribute to conservation and extant populations in the future. Since the first successful cryopreservation of bull semen (Polge *et al.*, 1949), cryopreserved bull semen has been used to propagate the rare and endangered species using assisted reproduction techniques.

The principle of testicular cell freezing and transplantation has been demonstrated and is currently used for human male infertility (Clouthier *et al.,* 1996). Significant efforts are being made on non-mammalian species using cryobiology techniques. In fish aquaculture, the successful cryopreservation of gametes and embryos could offer new commercial possibilities, allowing the unlimited production of fry and potentially healthier and better conditioned fish as required. Cryopreservation of reproductive products of many aquatic species has been successfully achieved. Cryopreservation of aquatic sperm is relatively common in the breeding and management of fish species, including salmonid, cyprinids, silurids, and Acipenseridae (família) is well documented. However, cryopreservation of embryos and oocytes of aquatic species has not been successful, except for eastern oyster eggs (*Crassostrea virginica*) (Tervit *et al.,* 2005), larvae of eastern oyster (Paniagua-Chavez and Tiersch 2001) and larvae of the sea urchin (Adams *et al.,* 2006).

Cryopreservation technology applied to the preservation of fish gametes in aquaculture plays an important role in seed production, genetic management of broodstock and conservation of aquatic resources. Fish germplasm also plays a significant role in human genomic studies because relatively small size of the genome makes it easier for sequencing and ideal models for studying the human disease. This would help in identifying the roles for human genes from fish mutations and also in fish models for genes identified by human disease. Aquatic species preservation would assist the development, protection and distribution of research lines and would offer benefits for restoration of endangered species.

Cryopreservation of Fish Sperm

First successful cryopreservation of fish sperm was reported in 1950s. The success of the cryopreservation of sperm depends on the seasonality, varying sperm quality, collection techniques, diluents used, precise freezing, thawing regimes, storage conditions, and post-thaw fertilization techniques.

The development of successful protocols for long-term freezing will allow the storage of disease-free gametes of desirable strains and species for future use and also it will facilitate in comparison and evaluation of new strains of fish with original parental stock at minimum cost. In addition, frozen gene banks may be used in conjunction with live gene banks. A major advantage of this approach is that the effective population size can be dramatically increased at relatively low cost by storing spermatozoa from a large number of individual males. Unlike sperm of higher vertebrates, fish spermatozoa remain quiescent and become activated from the moment when they come in contact with water. In most of the fishes spawning in freshwater has spermatozoa which remain motile for 2-3 min, and in carps energetic movement of sperm is only for a short duration of 30-60 secs during which they are capable to fertilize the eggs.

Cryopreservation methodology includes proper collection of fish milt; addition of extenders to prevent the depletion of sperm energy reserve and to maintain the sperm in quiescent condition but alive; by using cryoprotectants which reduces the thermal shock freezing and thawing techniques to minimize the sperm damage.

Many studies on cryopreservation of fish sperm have been carried out on economically important freshwater species and attempts to cryopreserve sperm from the marine fish species tended to be more successful when compared with those obtained from the freshwater fish (Tsvetkova *et al.,* 1996). Although freshwater fish sperm are generally more difficult to cryopreserve, the fertilization rates obtained from the cryopreserved marine fish sperm are similar to those obtained with mammalian species (Tsvetkova *et al.,* 1996). Controlled-rate slow cooling in cryopreservation has been mainly used for fish sperm. Common carp has been studied using frozen-thawed sperm with 95 per cent fertilization and hatching rate.

Salmonid species spermatozoa have been successfully cryopreserved (Lahnsteiner, 2000). Another well studied cryopreserved group is cyprinids and some of these cyprinid fishes are widely farmed throughout Asia and Europe. A fertilization and hatching rate of 95 per cent using the frozen-thawed sperm has been reported for the common carp and these results are not significantly different from fresh sperm (Magyary *et al.,* 1996). Tilapias are among the exotic freshwater fishes that have been successfully established for fish farming in Taiwan; they have been cryopreserved successfully and produced 40-80 per cent motility with cryoprotectant DMSO (Chao *et al.,* 1987). The sperm of more than 30 marine fish species have been cryopreserved successfully. Generally, high survival and fertilization capacity has been obtained in frozen-thawed spermatozoa when compared to freshwater species.

Dimethyl sulfoxide has also been reported as a successful cryoprotectant for sperm cryopreservation; the concentration range used was 5 to 30 per cent for these species. Various levels of motility, ranging from <5 per cent to 95 per cent, have been reported for the cryopreserved aquatic invertebrate sperm.

Materials Required

- Hank's Balanced Salt solution
- Cryoprotectants - Methanol
- Glycerol
- DMSO (10 per cent)
- French straws, Cryocan and its accessories

Procedure

- Collect the milt directly by pressing the abdomen of the matured male fish (in carps) or cut the testes and homogenize (in the case of catfish) in 10 ml HBSS (refrigerated at 4°C)

- Dissolve the cryoprotectants to HBSS or 0.9 per cent NaCl.
- Dilute the sperm suspension at 1:4 in various cryoprotectant solution.
- Draw the suspension into 0.5 ml French straws.
- Equilibrate for 15 min at 10°C prior to freezing.
- Place the straw in a strainer tray suspended above liquid nitrogen in a circular insulated tank.
- Check for sperm motility over a period of time.

Chapter 10

Protocol for Checking the Fish Sperm Motility

Introduction

Sperm motility is one of the most commonly used parameter to evaluate the sperm quality, sperm count must be more and motile to achieve fertilization. Unlike mammal, the teleost spermatozoa are found immotile in the testis as well as in seminal fluid. Healthy fish spermatozoa on activation shows a vigorous movement. The motility of such activated spermatozoa is evaluated by six point (++++++) scale after adding 100 times of water to a drop of milt.

Fish Sperm Motility

A simple observation of sperm under a microscope at a magnification between x10 and x25 using a glass slide with or without cover slip, has been so far extensively used in order to qualify spermatozoa properties before it is used in fertilization trials. This type of observation provides a coarse evaluation of sperm quality by motility criteria, since it allows only assessing classes in terms of per cent of motile sperm and motility duration defined by the time period leading to cessation of any progressive movement. Although an inaccurate and subjective, such a method highlighted the difficulties to objectively analyze the motility of spermatozoa and allowed to defining the bases for further individual analysis. A reliable motility assessment actually requires, a two-step dilution with a predilution in a non-activating medium and a high final dilution rate(>1/1000) in an activating medium adapted either to final sperm concentration pr motility efficiency in order to avoid heterogeneous triggering of motility.

Motility Percentage Score Condition

Percentage of Motile Sperm after Addition of Water	*Rating*	*Condition*
0 --10	+	All dead
11--30	++	Slightly active
31--50	+++	Oscillating
51--70	++++	Moving
71--90	+++++	Active
91--100	++++++	Excellent

Materials Required

Matured Male fish, Ringer' solution/physiological saline, Accessories for stripping milt, Haemocytometer, Microscope *etc.*

Procedure

1. Collect spermatozoa directly from the testis (for catfish and murrels). Otherwise strip the milt by pressing the abdomen gently (for carps).
2. Take 10 µl of sperm sample and add 490 µl of diluent Ringer' solution/ physiological saline in a micro centrifuge tube.
3. Draw 10 µl of the above diluted sample and add 90 µl of diluent (1:50 and 1: 10 dilution).
4. Transfer a drop of the milt content to a clean haemocytometer. Each square is divided into16 smaller cells (4x4). Each of these smaller cells has a volume of 1/4000 mm^3
5. Check the motility under a microscope after adding a drop of tap water and find out the percentage of motility
6. Active/excellent condition of motility of sperm is a minimum requirement for utilizing the fish for breeding.

Chapter 11

Problems on Heterosis in Fish Breeding

Introduction

It was observed that, when two pure strains or races (differing from each other in a number of genes) are crossed, the resultant hybrids may be markedly superior to either parent with regard to size, vigour, vitality, resistance to unfavorable environmental conditions and diseases, *etc.* This superiority of the hybrid was termed as heterosis by George Harrison Shull in the 1914. He along with Edward Murray East, conducted an elaborate studies of hybrid corn from 1905 to1930 and developed the genetic basis of heterosis concept. The exact genetic basis of heterosis is not properly understood. According to one view, growth, vigor, viability, fertility and other such "superior" qualities are governed by numerous genes, some of which are present in the dominant form in one pure race and others in the other pure race.

Heterosis

To clarify the genetic mechanism of heterosis, two hypotheses are usually considered; the hypothesis of combination in hybrids of favourable dominant factors, and the hypothesis of higher heterozygosity. The first, so called hypothesis of dominance (Jones, 1917), is based on a frequently observed correlation between dominance and favourable factors, as a result of overlapping of dominant genes over recessive genes in homologous chromosomes. The second hypothesis, that is, the idea of over-dominance, was originally proposed by G.H. Shull and E.M. East and definitely formulated by Hull (1945). According to this hypothesis, heterozygosis by itself promotes survival and development power.

From a genetic point of view it means that there are loci, the display of which proves to be more conspicuous in the heterozygous state than in the homozygous (Aa>AA>aa). In recent years most researchers admit the possibility of action of both mechanisms, while some of them believe that dominance and overdominance cannot be demarcated very strictly because of the presence of pseudoalleles and the genes which are located very close to each other and exert a similar influence on development. Moreover, it should be taken into consideration that, in the process of evolution, the significance of each of the proposed mechanisms does not remain unchanged and, in the long run, both the genetic mechanisms of heterosis can amalgamate into one.

Freshwater fishes has a number of biological peculiarities which facilitate the application of methods of commercial hybridization in pond fisheries. In addition, fishes manifest high fecundity, considerable magnitude of population, high level of natural heterozygosity, and explicit inbreeding depression in close breeding. The features determining higher heterozygosity of fish populations apparently condition the manifestation of heterosis in crossings of fishes. Taking these into account it appears possible to evolve a scheme of breeding best suited for obtaining and maintaining the maximum effect of heterosis under conditions of commercial hybridization.In all probability, heterosis of hybrids of freshwater fishes is due to overdominance. As experimental evidence of this, one can apparently consider demonstrated cases of heterosis on the molecular level in salmon hybrids, in whitefish hybrids (Kusakina, 1959, 1964) and in sunfishes (Manwell *et al.,* 1963).

Problem No. 1

A hatchery manager spawns and raises channel catfish, blue catfish, and then reciprocal hybrids in order to evaluate the relative growth of these catfishes at his hatchery. He harvest the four groups when they are 18 months old and records the following average weights. What is the heterosis in this experiment?

Group Avg. wt./g

Channel catfish 460 g

Blue catfish 440 g

Channel catfish ♀ x Blue catfish ♂ 600g

Blue catfish ♀ x Channel catfish ♂ 462g

Solution

$$\text{Heterosis} = \frac{\text{Avg. wt. of reciprocal } F_1 \text{ hybrids} \quad \text{Avg. wt. of parents}}{\text{Avg. wt. of parents}} \times 100$$

$$= \frac{531 \quad 450}{450} \times 100$$

= 18 per cent

Answer

Heterosis observed in the experiment was 18 per cent

Problem No. 2

A hatchery operator spawns and raises common carp, silver carp, and then reciprocal hybrids in order to evaluate the relative growth of these carps at his farm facility. He harvest the four groups when they are 12months old and records the following average weights. What is the heterosis in this experiment?

Group Avg. wt./kg

Common carp 1.200 kg

Silver carp 1.600 kg

Common carp ♀ x silver carp ♂ 1.750 kg

Silver carp ♀ x common carp ♂ 1.450 kg

Solution:

$$\text{Heterosis}=\frac{\text{Avg. wt. of reciprocal } F_1 \text{ hybrids} \quad \text{Avg. wt. of parents}}{\text{Avg. wt. of parents}}\times 100$$

$$=\frac{1.600 \quad 1.400}{1.400}\times 100$$

= 14.28 per cent

Answers

Heterosis observed in the experiment was 14.28 per cent

Problem No. 3

A hatchery operator produces fry and raises magur, African catfish, and then reciprocal hybrids in order to evaluate the relative growth of these catfishes at his experimental facility. He harvest the four groups when they are 10 months old and records the following average weights. What is the heterosis in this experiment?

Group Avg. wt./g

Magur 800 g

African catfish 1400 g

Magur ♀ x African catfish ♂ 1350 g

African catfish ♀ x Magur ♂ 1250 g

Solution

$$\text{Heterosis}=\frac{\text{Avg. wt. of reciprocal } F_1 \text{ hybrids} - \text{Avg. wt. of parents}}{\text{Avg. wt. of parents}}\times 100$$

$$= \frac{1300 \quad 1100}{1100} \times 100$$

= 18.18 per cent

Answers

Heterosis observed in the experiment was 18.18 per cent

Chapter 12

Problems on Sex Determination

Introduction

A sex determination system is a biological system that determines the development of sexual characteristics in an organism. Most sexual organisms have two sexes. In many cases, sex determination is genetic: males and females have different alleles or even different genes that specify their sexual morphology. In animals, this is often accompanied by chromosomal differences. In other cases, sex is determined by environmental variables (such as temperature) or social variables (the size of an organism relative to other members of its population).

Physiological Sex

Physiological sex is formed through the biochemical process of onto genesis under the control of genetic sex.

Gonadal Sex

The basis of the physiological sex depends upon the type of primary sex organ. This basic sex is known as the gonadal sex which is further classified into two groups, gonochorism and hermaphroditism. Gonochorism type is the most common type in fishes.

Sex Determination Mediated by Sex Chromosomes

The determination of sex by chromosomes differing in one or few of the genes is a characteristic features of many species. For example, salmonidae and cyprinidae. The genetic factors affecting the sex (acting apparently via the hormones determining the sex) are, present in autosomes as well. Nevertheless, the main part is played by the sex genes located in the sex chromosomes.The identification of sex chromosomes in fishes is a difficult task. The sex chromosomes are distinguished from other chromosomes by the presence of sex-chromatin (barr body) and karyotypic difference between two sexes in somatic metaphase counts.

This type of sex determination is of more advanced type. The sexual diversity also occurs at the level of chromosomes. Within this mechanism, the sex chromosomes X and Y or W and Z are different, but the difference merely involve the existence of one or several specific male and female sex genes.In *Poecilia reticulata, P. variatus, Oryzias* sp. *Salmo gairdneri* and *Xiphophorus maculatus* distinct sex chromosome has been reported. Sex chromosomes have been marked using the genes responsible for pigmentation.

In many fishes, female has two X-chromosomes, male has only one. All X-chromosome genes in the female are present in a double dose and in male single dose. This difference in dosage would have an effect on gene expression, leading to unbalance in one of the two sexes. Male heterogamety XY was most common.

XX♀:XY♂– System

- Most common system found even in Man. XX female (homogametic), XY male (heterogametic).
- This type of system is reported in channel catfish, common carp, rainbow trout, sockeye salmon (*Oncorhynchusnerka*), coho salmon (*O. kisutch*), goldfish, nile tilapia, grass carp, murrels, *Oryziaslatipes, Poeciliareticulata, etc.*
- The fact that XXY triploid rainbow trout are males suggests that a "dominant Y" sex-determining mechanism is operative.

ZZ♂:WZ♀– System

- To avoid confusion XY system is omitted.
- ZZ male (homogametic), WZ females (heterogametic).
- This system is reported in fishes such as *T. aurea, T. hornorum, Gambusia affinis* (mosquito fish), Mud skipper (*Boleophthalmus boddaerti*), *Aplocheilus panchax*, European eel (*Anguilla anguilla*), Japanese eel (*A. japonica*), freshwater prawn (*Macrobrachium rosenbergii*).

WXY – System

- This system is reported in platy (*Xiphophorus maculatus*).
- Y- chromosome produces males except when it is paired with a W chromosome.
- The W chromosome is a modified X chromosome that blocks the male determining function of the Y chromosome.
- There are three sex chromosomes in *X. maculatus*, W, X, and Y.

- ☆ Three combinations of sex chromosomes, WY, WX and XX cause female differentiation, and two combinations, XY and YY give rise to males.
- ☆ In wild population, six kinds of matings occur of which four result in an even sex ratio, one gives rise to a ratio of 3 females to 1 male, and one yields all male progeny.

XO – System and ZO – System

- ☆ XO – system. "O" is the symbol for no chromosome.

  XX are females and XO are males.

  E.g. Sternoptyxdiaphana, Oncorhynchusnerka, Corisjulis

- ☆ ZO-system. ZZ males, and ZO females.

  E.g. Colisalalius, C. fasciatus

Multiple Sex Chromosomes

- ☆ $X_1 X_2 X_2$ – females, $X_1 X_2$ Y– males.

  E.g. Megupsilonaporus (File fish), *Cyprinodon* sp.

- ☆ ZZ/ZW_1 W_2. Multiple W chromosomes are present. ZZ–males, ZW_1 W_2 – females

  E.g. Apareiodon affinis

- ☆ X$Y_1 Y_2$ males, XX – females.

  E.g. Hoplias sp.

  In this system the number of chromosomes is not constant within a species.

  In the first two cases, females have one extra chromosomes. In the third, males have one extra chromosomes.

  Sex determination in fish that have sex chromosomes is not really a simple. Although sex in these species is controlled by the sex chromosomes, an individual's sex can also be influenced or controlled by autosomal sex-influencing or sex-modifying genes (as in the case of tilapia).

Problem No. 1

The presence of different types of sex chromosomes allows different sex ratios to be obtained in various crosses in the case of tilapia. Find out the sex ratios for the following crosses.

1. XX × XY
2. ZZ x WZ
3. WZ (sex reversed) × XY
4. XX x ZZ

Solution

1. XX x XY

Parent XX x XY

F1

	X	Y
X	XX	XY
X	XX	XY

The ratio of XX: XY = 1: 1

1. ZZ x WZ

Parent ZZ x WZ

F1

	W	Z
Z	WZ	ZZ
Z	WZ	ZZ

The ratio of ZZ: WZ = 1: 1

- WZ × XY
  (sex reversed)
  Parent WZ × XY
  F1

	X	Y
W	WX	WY
Z	ZX	WZ

The ratio of WX: ZX: WY: ZY = 1: 1:1:1

- XX x ZZ
  Parent XX x ZZ
  F1

	Z	Z
X	XZ	XZ
X	XZ	XZ

Answers

a) The ratio of XX: XY = 1: 1
b) The ratio of ZZ: WZ = 1: 1
c) The ratio of WX: ZX: WY: ZY = 1: 1:1:1
d) All XZ

Problem No. 2

Find out the sex ratio of F_1 hybrid progeny from a cross of a female*Oreochromis mossambicus* and a male *O. hornorum*?

Solution

The Sex chromosomes for *O. mossambicus* female - XX

Sex chromosomes for *O. hornorum* male – ZZ

Parent XX x ZZ

F1

	X	X
Z	XZ	XZ
Z	XZ	XZ

Answer

The sex ratio of F_1 hybrid progeny from a cross of a female *Oreochromis mossambicus* and a male *O. hornorum* is all males (XZ)

Problem No. 3

Find out the genotypic and phenotypic ratio for the mating of a female grey and maculatus male guppy?

Solution

The genotype of female grey guppy is XX and the genotype of maculates male guppy is XYma

Parent XX x XY ma

F1

	X	Yma
X	XX	XYma
X	XX	XYma

Answer

Genotypic ratio - 1 XX: 1 XY ma

Phenotypic ratio - 1 grey female; 1 maculatus male

Problem No. 4

In platy, a cross between a 'Nigra' female (N) and a lightly pigmented male (W) produced 'Nigra' sons and 'white' daughters which, in an F_2, produced approximately equal frequencies of 'Nigra' and 'white' individuals of each sex. Interpret the results of the above cross?

Solution

Parent WO Zn X ZwZw

Nigra ♀ White ♂

F1

	Wo	Zn
Zw	WoZw	Zn Zw
Zw	WoZw	Zn Zw

WoZw : Zn Zw = 1: 1

Answers

WoZw: Zn Zw = 1: 1

White ♀: Nigra ♂= 1: 1

Problem No. 5

Find out the sex ratio for the progeny of the following crosses produced in guppy?

- ☆ Caudalis ♀ × caudalis ♂
- ☆ Transparent tailed ♀ × caudalis ♂
- ☆ Caudalis ♀ × transparent tailed ♂

Solution

a) caudalis ♀× caudalis ♂

ParentXcpXcp × XcpY

	Xcp	Y
Xcp	XcpXcp	XcpY
Xcp	XcpXcp	XcpY

b) transparent tailed ♀× caudalis ♂

ParentXchXch × XcpY

	Xcp	Y
Xch	XcpXch	XchY
Xch	XcpXch	XchY

c) caudalis ♀× transparent tailed ♂

Parent

XcpXch × XchY

	Xch	Y
Xcp	XcpXch	XcpY
Xch	XchXch	XchY

Answers

- ☆ Genotypic ratio - 1 XcpXcp: 1 XcpXcp: 1 XcpY: 1 XchY

  Phenotypic ratio - 2 caudalisfemale: 1 caudalis male:1 transparent tailed male.
- ☆ Genotypic ratio - 1 XcpXcp: 1 XchY

  Phenotypic ratio - 1 caudalis♀: 1 transparent tailed ♂
- ☆ Genotypic ratio - 1 XchXch: 1 Xcp Xch: 1 Xch Y: 1 Xcp Y

  Phenotypic ratio - 1 transparent tailed ♀: 1 caudalis ♀: 1 transparenttailed ♂: 1 caudalis ♂

Problem No. 6

In the case of Medaka, *Oryziaslatipes* the mechanism of sex determination is XY system. Find out the sex ratio among the progeny produced from the following crosses.

- ☆ XX × XX (T)
- ☆ XY (T) × XY
- ☆ XY (T) × YY (from the second cross)
- ☆ YY (T) × YY (from the second cross)
- ☆ XX × YY (from the second cross)

Solution

a) XX × XX (T)

Parent XX × XX (T)

	X	X (T)
X	XX	XX
X	XX	XX

b) XY (T) × XY

ParentXY (T) × XY

	X	Y
X	XX	XY
Y (T)	XY	YY

c) XY (T) × YY (from the second cross)

Parent XY (T) × YY

	Y	Y
X	XY	XY
Y	YY	YY

d) YY (T) × YY (from the second cross)

Parent YY (T) × YY

	Y	Y
Y	YY	YY
Y (T)	YY	YY

e) XX × YY (from the second cross)

Parent XX × YY

	Y	Y
X	XY	XY
X	XY	XY

Answers

- ☆ All females
- ☆ 3 males: 1 female
- ☆ All males
- ☆ All males
- ☆ All males

Chapter 13

Problems on Inbreeding and Effective Breeding Number

Introduction

Inbreeding is also one of the breeding programme that can have a tremendous impact on productivity. As recently 200 years ago there were laws for bidding consanguineous matings even in livestock, because it was considered immoral and against the laws of God and Nature. But according to plant breeders, inbreeding is one of the most important breeding techniques; without its use, agricultural productivity would decline.

Inbreeding

Inbreeding is simply a mating of related individuals. Inbreeding does not imply, nor does the definition mention, anything about viability, growth, or productivity. (A mating between relatives is often referred to as consanguineous mating). Inbreeding is neither good nor bad. Genetically, all inbreeding **increases homozygosity** in the offspring (this means inbreeding also **decreases heterozygosity** in the offspring by an equal amount). The increase in homozygosity occurs because related fish share alleles through one or more common ancestors, *i.e.* the parents may carry a copy of an allele that both inherited from a common ancestor.

When the relatives mate, the alleles that they share because of their common ancestor(s) can be paired in their offspring. This produces offspring that are more likely to be homozygous at one or more loci. The mating of unrelated fish also produces offspring that have homozygous genes. Additionally, an inbred fish looks the same as one with no inbreeding; there is no distinguishing mark that separates fish into inbred vs non-inbred categories.

This increase in homozygosity is called "inbreeding", and the coefficient of inbreeding is a measure of how much more homozygous fish is in that population at an average. The coefficient of inbreeding does not measure how many homozygous loci the fish has; it simply quantifies the per cent increase in homozygosity. On a gene-by-gene basis, **F** is the probability that the two alleles will be identical by descent. Since **F** measures the per cent increase in homozygosity, the same level of inbreeding can produces a different amount of homozygosity in different populations, depending on the level of homozygosity that has occurred over a specific time interval (number of generations) due to identity by descent (due to the mating of relatives). Additionally, F is based on probability, so a given value of F is an average value; consequently, a given value of F means more homozygosity will have been produced in some fish, while less will have been produced in others.

Effective Breeding Number (N_e)

The N_e of a hatchery population is one of the most important bits of information about the population. Unfortunately, in most hatchery managers do not know what N_e is, do not know how to determine N_e, and do not know how N_e influences inbreeding and brood stock management. If you asked to describe the size of their hatchery population, most farmers or hatchery managers would produce a census or an approximate number. Those with good records would probably be able to give the number of male and female brood fish. This information is used to determine the number of fingerlings that can be produced, to calculate the amount of feed that must be ordered, or to estimate yield. As important as this information is, it does not describe the population genetically. In order to describe a population genetically, one must determine N_e.

Effective breeding number is one of the most important concepts in brood stock management, because it gives an indication about the genetic stability or genetic health of the population. This is because N_e is inversely related to inbreeding and to genetic drift. If hatchery populations were infinitely large, an understanding of N_e would be unnecessary. However, hatchery populations are usually small and are often closed. A closed population is one where immigration (the introduction of fish from another population) is not allowed; consequently, fish from other populations are not allowed to mate with or hybridize with fish from a closed population. Hatchery managers often maintain closed populations for various reasons; chief among them is the desire to the minimize health problems by preventing the introduction of disease that often accompany a acquired fish.

When working with closed, finite population, the best way to describe it is not by total number of fish, but by N_e. Effective breeding number is determined by the number of male and female brood fish that produce viable offspring, the sex ratio of the brood fish that spawned, the variance of family size, and the mating system that is used. In most situations where fish cannot be identified and where mating is random (fish are paired without regard to phenotypic value, or fish swim free in a pond and choose their own mates), N_e can be determined by using the following formula:

$$N_e = \frac{4(\text{Number of males})\,(\text{Number of females})}{\text{Number of males} + \text{Number of females}}$$

where, number of males and females are the number of male and female brood fish that produce viable offspring. If all offspring for a brood fish die, that male or female is not included while determining N_e.

If the matings and offspring production cannot be monitored, N_e cannot be determined. If fish spawn in ponds and eggs are allowed to hatch in the ponds and if offspring are not harvested until they are mixed schools of fry or fingerlings, N_e will be difficult, if not impossible, to determine.The formula shows that N_e is determined by both the number of male and female brood fish and by the sex ratio. Effective breeding number and the number of brood fish that produce offspring will be same only when the sex ratio is 1:1.

Once N_e has been determined, a simple formula can be used to calculate the average inbreeding value in the population:

$$F = \frac{1}{2N_e}$$

where, F is the average inbreeding value in the population

Problem No. 1

Calculate the effective breeding number (Ne), if 53 females and 25 males were used for breeding. What will be the effective breeding number if the sex ratio is 1:1.

Solution

No. of female= 53

No. of male = 25

$$N_e = \frac{4(\text{Number of males})\,(\text{Number of females})}{\text{Number of males} + \text{Number of females}}$$

$$= \frac{4 \times 25 \times 53}{25 + 53}$$

$$= \frac{5300}{78}$$

= 67.94

If the sex ratio is 1: 1

No. of female= 53

No. of male = 53

$$N_e = \frac{4(\text{Number of males})\,(\text{Number of females})}{\text{Number of males} + \text{Number of females}}$$

$$= \frac{4 \times 53 \times 53}{53 + 53}$$

$$= \frac{11236}{106}$$

= 106

No. of female = 25

No. of male = 25

$$N_e = \frac{4(\text{Number of males})\,(\text{Number of females})}{\text{Number of males} + \text{Number of females}}$$

$$= \frac{4 \times 25 \times 25}{25 + 25}$$

$$= \frac{2500}{50}$$

= 50

Answers

The effective breeding number (Ne), if 53 females and 25 males were used for breeding is 67.94. The effective breeding number if the sex ratio is 1:1 either 106 or 25.

Problem No. 2

In an closed isolated natural fish population, the number of breeding pairs is estimated to be1050 females and 950 males. Calculate the effective breeding number and the rate of inbreeding.

Solution

$$N_e = \frac{4(\text{Number of males})\,(\text{Number of females})}{\text{Number of males} + \text{Number of females}}$$

$$= \frac{4 \times 950 \times 1050}{950 + 1050}$$

$$= \frac{3990000}{2000}$$

= 1995

$= \frac{1}{2N_e}$

$= \frac{1}{2 \times 1995}$

= 0.00025

Answers

The effective breeding number and the rate of inbreeding for the given population are 1995 and 0.00025 respectively.

Problem No. 3

Find out the effective breeding number for a selective breeding programme, if you wish to decide 5 per cent as the critical level for inbreeding for next 15 years.

Solution

Step 1: Calculated the F/generation needed to produce F=0.05 at 15 generations.

F/generation = 0.05/15

= 0.0033333

Step 2: Calculate the $= \frac{1}{2N_e}$

Ne = 1/(0.000333330 (2)

= 1.150

Solution

The effective breeding number to be followed for each generation is 150

Chapter 14

Problems on Selection and Heritability

Introduction

The branch of genetics dealing with the genetic model for quantitative traits and its applications is called quantitative genetics.Quantitative phenotypes are those that are measured, such as length, weight, and fecundity. The important production phenotypes are quantitative ones, although some qualitative phenotypes are quite important, and can greatly increase the value of a crop.

Because quantitative phenotypes can be measured, each phenotype is a single category, such as length. Individuals do not get segregated into alternate phenotypic categories, such as long vs short instead, they are distributed along a continuum, and differences among individuals are determined by the unit of measure that is used to assess the phenotype; millimetres, grams, *etc.* Because each fish's phenotype is determined by a measurement, quantitative phenotypes form are called as "continuous distributions", which are described by the population's mean and the distribution about the mean (variance and standard division).

In a population, these phenotypes form are called as "normal" or "bell-shaped" distributions. The reason why quantitative phenotypes exhibit continuous distributions and why individuals do not get segregated into descriptive categories is that quantitative phenotypes are far more complicated genetically than qualitative phenotypes. Each quantitative phenotype is controlled by dozens to hundreds of genes, each of which makes a small contribution to the production of the phenotype. The exact number of gene is usually never known.

Phenotypic Variance (V_P)

Phenotypic variance (V_P) is the sum of the genetic variance (V_G), environmental variance (V_E), and genetic-environmental interaction variance (V_{G-E}) components.

$$V_P = V_G + V_E + V_{G-E}$$

When conducting a breeding programme, a geneticist tries to exploit $\mathbf{V_G}$. Three distinct types of genetic variance combine to make $\mathbf{V_G}$, and it is important to know what they are, because different breeding programmes are needed to exploit each type.

Genetic Variance

Genetic variance is the sum of additive genetic variance (V_A), dominance genetic variance (V_D), and epistatic genetic variance (V_I):

$$V_G = V_A + V_D + V_I$$

These components of genetic variance do not refer to additive, dominance, and epistatic gene action; they refer to specific components of phenotypic variance that are produced by the entire genome, not that produced by one or two genes. The major components are V_A and V_D. Epistatic genetic variance is usually considered to be unimportant because it is difficult to exploit and improvements that occur by exploiting V_I plateau quickly.

Additive Genetic Variance

Additive genetic variance is the component that is due to the additive effect of all the fish's alleles taken independently *i.e.*, the sum of the effects that each allele makes to the production of phenotype. Additive genetic variance is the most important component of V_P, and the percentage of V_P, which is controlled by V_A is called "heritability" (h2).

$$h^2 = V_A/V_P$$

1. Additive genetic variance is the genetic component that can be exploited by selective breeding programmes.
2. The breeding programme needed to exploit V_D is cross breeding. When h2 is ≤ 0.15, crossbreeding is often prescribed to exploit V_G.
3. The major reason for determining the heritability of a quantitative phenotype is that it can be used to predict the results of a selective breeding programme by using the following formula,

   $$R = S\,h^2$$

   Where **R** is the response to selection (gain per generation), **S** is the selection differential (the superiority of the select broodstock over the population average) and $\mathbf{h^2}$ is heritability.

Problem No. 1

A catfish farmer decides to initiate a selection programme for increased growth rate in the marion strain of channel catfish, which currently averages 454 g at 18 months of age at his farm. To implement his programme, the farmer selects 50 females that average 604g and 40 males that average 692g. What is the predicted average weight in the next generation?

Given: h^2 for growth in this strain is 0.50

Solution:

Step 1. h2 for growth in this strain is 0.50 (Dunham and Smither-man 1983).

Step 2. The selection differential(S) is calculated as follows:

S = [(mean wt.selected ♀ + mean wt.selected ♂)/2] – population mean

S = [(604 g + 692 g)/2] – 454 g

S = 194 g

Step 3. Calculate the predicted response to selection R = Sh2

R = (194 g) (0.5)

R = 97 g

The next generation of Marion channel catfish at the hatchery (F_1) should average 97 g more than the parental generation (P_1) and will average

$F_1 = P_1$ mean weight + response to selection

$F_1 = 454$ g + 97 g

$F_1 = 551$ g

Answer

The next generation of Marion channel catfish at the hatchery (F_1 will average 551 g.

Problem No. 2

How to predict the response to selection if the heritability (h2) of the phenotype is known. Predict response to selection for increased length in common carp and then calculate the predicted mean length of the next generation.

Given: h2 for length at 12 months = 0.26

Mean length at 12 months of the population = 146 mm

Mean length at 12 months of the select brood fish = 162 mm

Solution

Step 1. h2 for growth in this strain is 0.26.

Step 2. The selection differential(S) is calculated as follows:

S = [(mean length selected ♀ + mean length selected ♂)/2] – population mean

S = [(162 mm + 162 mm)/2] - 146mm

S = 16 mm.

Step 3. Calculate the predicted response to selection R = Sh2

R = (16 g) (0.26)

R= 4.16 mm

The next generation of the hatchery (F_1) should average 4.16 mm in length more than the parental generation (P_1) and will average

F_1 = P_1 mean length + response to selection

F_1 = 146 mm + 4.16 mm

F_1 = 150.16 mm.

Answers

The next generation of the hatchery (F_1) will average150.16 mm.

Problem No. 3

A tilapia farmer decides to initiate a selection programme for increased growth rate, which currently averages 400 g at 6 months of age at his farm. To implement his programme, the farmer selects 100 females that average 650 g and 140 males that average 550 g. What is the predicted average weight in the next generation?

Given: h^2 for growth in this strain is 0.36.

Solution

Step 1. h2 for growth in this strain is 0.36

Step 2. The selection differential(S) is calculated as follows:

S = [(mean wt.selected ♀ + mean wt.selected ♂)/2] – population mean

S = [(650 g + 550 g)/2] – 400 g

S = 200 g

Step 3. Calculate the predicted response to selection R = Sh2

R = (200 g) (0.36)

R=72 g

The next generation of tilapia at the hatchery (F_1) should average 72 g more than the parental generation (P_1) and will average

F_1 = P_1 mean weight + response to selection

$F_1 = 400 \text{ g} + 72 \text{ g}$

$F_1 = 472 \text{ g}$

Answers

The next generation of tilapia at the hatchery (F_1) will average472 g.

Chapter 15

Problems on Gene and Genotype Frequency

Introduction

Allele frequency, or gene frequency, is the relative frequency of an allele (variant of a gene) at a particular locus in a population, expressed as a fraction or percentage. Specifically, it is the fraction of all chromosomes in the population that carry that allele. Genotypic frequency is the frequency of a genotype – homozygous recessive, homozygous dominant, or heterozygous – in a population.

Problem No. 1

A trout farmer finds some golden and palomino rainbow trout among his normally pigmented rainbow trout at his farm and wants to know the frequencies of the alleles responsible for the production of these phenotypes. A census of his population shows the following:

Phenotype	Genotype	Number
Normally pigmented	GG	360
Golden	G'G'	160
Palomino	GG'	480
	Total	1000

Also find out the genotypic frequency for the three phenotypes based on the gene frequency estimated.

Solution

☆ Calculation of allele frequency:

$$f(G) = \frac{\text{No.of G alleles}}{\text{Total No.of alleles at G locus}}$$

$$f(G) = \frac{2(360) + 480}{2(1000)}$$

$$f(G) = \frac{1200}{2000}$$

= 0.6

$$f(G) = \frac{\text{No.of G alleles}}{\text{Total No.of alleles at G locus}}$$

$$f(G) = \frac{2(160) + 480}{2(1000)}$$

$$f(G) = \frac{800}{2000}$$

= 0.4

Note that f (G) +f(G')= 0.6+0.4=1

Calculation of Genotypic Frequency

	G (f=0.6)	G' (f=0.4)
G (f=0.6)	f (GG) = 0.6 x0.6 =0.36 Normally pigmented	f (GG') = 0.6 x0.4 =0.24 Palomino
G' (f=0.4)	f (G'G) = 0.4 x0.6 =0.24 Palomino	f (G'G') = 0.4x0.4 =0.16 Golden

Answers

Frequency of G allele=0.6

Frequency of G' allele=0.4

Genotypic frequency: The genotypic frequency of normally pigmented, Palomino and golden rainbow trout are 0.36,0.48 and 0.16 respectively.

Problem No. 2

A carp farmer finds some blue common carp at his hatchery. The census of his population shows the following:

Phenotype	*Genotype*	*Number*
Blue	bb	90
Normally pigmented	BB/Bb	910

Solution

$$f(b) = \sqrt{f\,(\text{blue common carp})}$$

$$f(b) = \sqrt{90/1000}$$

$$= 0.3$$

$$f(B) = 1.0\text{-}0.3$$

$$= 0.7$$

Calculation of Genotypic Frequency

	B(f=0.7)	b(f=0.3)
B(f=0.7)	f (BB) = 0.7 x0.7 =0.49 Normally pigmented	f (Bb) = 0.7 x0.3 =0.21 Normally pigmented
b (f=0.3)	f (Bb) = 0.3 x0.7 =0.21 Normally pigmented	f (bb) = 0.3x0.9 =0.09 Blue

Answers

Frequency of B allele=0.7

Frequency of b allele=0.3

Genotypic frequency: The genotypic frequency of normally pigmented (BB), normally pigmented (Bb) and blue

Common carp are 0.49, 0.42 and 0.09 respectively.

Problem No. 3

A guppy farmer drained a pond and found the following:

Phenotype	*Genotype*	*Number*
Grey females	XX	1650
Maculatus males	XY_{Ma}	340
Grey males	XY	510

Calculate the frequency of Y_{Ma} gene in this population.

Solution

$$f(Y_{Ma}) = \frac{\text{Maculatus males}}{\text{Total number of males}}$$

$$= \frac{340}{850}$$

$$= 0.4$$

Answer

The frequency of YMa gene in this population is 0.4.

www.ingramcontent.com/pod-product-compliance
Ingram Content Group UK Ltd.
Pitfield, Milton Keynes, MK11 3LW, UK
UKHW021955270726
14060UKWH00002B/534